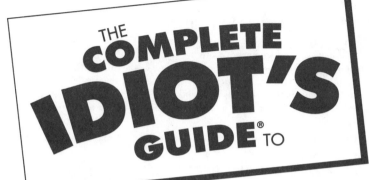

# THE COMPLETE IDIOT'S GUIDE® TO

# Science Fair Projects

*by Nancy K. O'Leary and Susan Shelly*

## ALPHA

A member of Penguin Group (USA) Inc.

## Copyright © 2003 by Nancy K. O'Leary and Susan Shelly

International Standard Book Number: 1-59257-137-9
Library of Congress Catalog Card Number: 2003111797

05  04  03      8  7  6  5  4  3  2  1

Interpretation of the printing code: The rightmost number of the first series of numbers is the year of the book's printing; the rightmost number of the second series of numbers is the number of the book's printing. For example, a printing code of 03-1 shows that the first printing occurred in 2003.

*Printed in the United States of America*

**Note:** This publication contains the opinions and ideas of its authors. It is intended to provide helpful and informative material on the subject matter covered. It is sold with the understanding that the authors and publisher are not engaged in rendering professional services in the book. If the reader requires personal assistance or advice, a competent professional should be consulted.

The authors and publisher specifically disclaim any responsibility for any liability, loss, or risk, personal or otherwise, which is incurred as a consequence, directly or indirectly, of the use and application of any of the contents of this book.

**Publisher:** *Marie Butler-Knight*
**Product Manager:** *Phil Kitchel*
**Senior Managing Editor:** *Jennifer Chisholm*
**Senior Acquisitions Editor:** *Randy Ladenheim-Gil*
**Senior Development Editor:** *Michael Thomas*
**Senior Production Editor:** *Christy Wagner*
**Production Editor:** *Megan Douglass*
**Copy Editor:** *Cari Luna*
**Illustrator:** *Chris Eliopoulos*
**Additional illustrations by** *Niemczyk Hoffman Group*
**Cover/Book Designer:** *Trina Wurst*
**Indexer:** *Aamir Burki*
**Layout/Proofreading:** *Mary Hunt and Ayanna Lacey*

# Contents at a Glance

# Contents

# Foreword

By freshman year of high school I suspected that I was going to be a scientist. Much to the dismay of my parents, half of our basement was occupied with a home laboratory. Foul smells and blown electrical circuits were frequent household events.

I was genuinely excited by the forthcoming science fair, but many of my friends didn't share this enthusiasm. Facing the same major decision as my schoolmates, I became confused and science became complicated. What in the world was I going to do? There were various kinds of electronic devices and assorted gadgets in the basement lab, but I needed an experiment and I desperately wanted a *special* experiment. Much like the cartoon light bulb turning on, I suddenly had an unpleasant thought. I didn't know how to conduct a proper experiment.

Many trips to many libraries solved this problem as the scientific method was revealed to me. I won first place awards at regional and state level during the next four years before college. There is no doubt these science fairs helped to shape my future. I also found that I became a ready resource for friends at school. This hasn't changed. I still help students all over the world through the Internet and I still enjoy it. The same methodology I used in high school science fair projects helped me produce my doctoral dissertation and all my other research.

We live in a marvelous and exciting world of wonders. There is so much to explore, but where do you start? You students are still confronted with the same dilemma that I was, and your parents and teachers can use all the help they can get, too! What will you do and how will you do it? What rules and guidelines do you need to follow? And those questions are only the beginning!

I have read many "how to" books on science fairs and I was delighted with *The Complete Idiot's Guide to Science Fair Projects* by Nancy K. O'Leary and Susan Shelly. This book is easy to read, and right on target. The first six chapters introduce you to the science fair system, coach you on how to find a topic, and most important, describe the scientific method and how to apply it. You will find many "Scientific Surprises" that explain scientific events and experiments as well as interesting stories of science history.

The second part of the book is loaded with cool and scientifically significant experiments from several disciplines that are designed to get you thinking. These experiments start at a basic level, move to intermediate and proceed to advanced level work. Each experiment is an excellent learning experience with complete material lists, procedures, and explanations. You are also presented with useful "Explosion Ahead" notes to caution you on an experimental procedure. I tried one of the experiments and learned how to extract and compare DNA strands. It worked on the first try with fascinating results. This is real science made readable, doable, and exciting.

If you are a student or parent trying to develop an interesting topic and a solid scientific experiment for the science fair, *The Complete Idiot's Guide to Science Fair Projects* is a terrific place to start. If you are a science teacher looking for exciting and significant experiments, you will find them in this book. Read this book and you will discover the fun in scientific experimentation and begin to think and work like a scientist.

—John W. Gudenas, Ph.D.

John W. Gudenas holds a Ph.D. and M.S. in Computer Science and a B.S. in Physics. He has worked at Argonne National Laboratory and is currently a Professor of Computer Science. Dr. Gudenas has published numerous articles and is the founder and director of "Ask Dr. John," the critically acclaimed Internet help feature for science fairs. You can find "Ask Dr. John" at www.SciFair.org.

# Introduction

Some folks love doing science fair projects, while others dread them. Chances are that you fall somewhere in the middle of that continuum.

In this book, we've presented a wide variety of projects suitable for different age levels and interests. It's no fun, and not a good idea, to undertake a project in which you have no interest. Your lack of enthusiasm and appreciation for the project will be evident in your work, resulting in a mediocre job, at best.

When you select a topic that excites you, however, a project becomes more than an assignment. It becomes an opportunity for you to challenge yourself, to be creative, and to stretch your limits.

The experiments in this book are presented in a clear, easy-to-understand format, and are fun to do. Choose one that excites you, and anticipate a great science fair project.

## What You'll Find in This Book

*The Complete Idiot's Guide to Science Fair Projects* is written in six sections. Four of those sections present science fair projects, while the other two cover other topics relating to science fairs.

**Part 1, "So You're Going to Do a Science Fair Project,"** starts at the beginning by explaining exactly what a science fair is. It tells you about different types of fairs, and includes plenty of good recommendations for choosing a topic.

The scientific method is an integral part of conducting an experiment, and you'll learn all you need to know about it in this section. You'll also learn the fine points of collecting data and making observations during the course of your project.

**Part 2, "Science Projects for Beginners,"** contains five chapters filled with great ideas for exciting projects for younger kids.

Each chapter contains a project that is thoroughly explained, taking you through the experiment step-by-step and providing charts on which you can record your results. In addition, each chapter contains two additional ideas for projects, with ideas on how to get them started.

**Part 3, "Intermediate-Level Science Projects,"** raises the bar for the project level, but provides the same thorough explanations and aids.

Students who choose a topic from this section may experiment with and learn about windchill, reaction time, static electricity, and other fun subjects.

The caliber of projects is raised again in **Part 4, "Advanced-Level Science Projects,"** intended, of course, for older students.

Here, students can choose a topic that explores animal DNA, plant cloning, corrosion of metals, or other topical, interesting areas of science. As with the two preceding sections, each chapter contains a thorough explanation of one project, and outlines ideas for two other projects relating to the same area of science.

More projects are presented in **Part 5, "Projects to Really Wow 'Em."** These are fun projects, dealing with offbeat topics such as exploding sandwich bags, the amount of food energy found in different kinds of nuts, and balloon rockets. You can even learn how to make your own lava lamp.

**Part 6, "Preparing Your Project for the Science Fair,"** covers the nuts and bolts of building a display, offering lots of suggestions concerning creativity, neatness, and practical issues like transporting it.

You'll learn exactly what the judges are looking for in a project, and how you can make your best impression. The final chapter will even get you thinking about next year's fair.

When you finish the book, you'll have lots of good ideas for projects, as well as a sound understanding of the mechanics of a science fair. You won't be an expert, mind you, but you'll be on your way.

## More Bang for Your Buck

*The Complete Idiot's Guide to Science Fair Projects* includes four types of sidebars. These little bits of information are intended to provide warnings to keep you out of trouble; give you practical, usable tips; tell you what words or phrases mean; or provide interesting snippets of information that you can tell your friends.

**Standard Procedure**

These tips provide additional ideas and help you sail through your project.

**Basic Elements**

Scientific lingo can be a little daunting, but these sidebars help you understand the talk.

**CAUTION**

**Explosion Ahead**

Don't even go there! These sidebars help you avoid trouble—even disaster—during the course of your project.

**Scientific Surprise**

There are millions of interesting little facts about science. Some of them are presented in these sidebars.

## Acknowledgments

The authors would like to thank the many people who provided time, information, and resources for this book. Especially, we want to thank our editors at Alpha Books, Randy Ladenheim-Gil, Michael Thomas, and Christy Wagner, for their direction and enduring patience. Many thanks also to Gene Brissie of James Peter Associates for his guidance and insights.

Nancy O'Leary would like to thank her patient and loving family for allowing her the time to pursue this project. Special thanks to her son, Joe, for his permission to include one of his blue-ribbon science fair projects in this book. To her daughter, Jackie, for her assistance in testing some of the experiments and for her violin and guitar serenades. And, to her husband, John, for his great support and encouragement throughout the past, busy months.

She also would like to thank her friend and neighbor, Melanie, for permission to include her award-winning science fair project.

Susan Shelly would like to thank Joan and Emma Jean Morse for their great ideas; Wendy Merz for technical assistance and Hot Pockets tips; and her special husband, Michael, and kids, Sara and Ryan, for keeping the office door shut when she was working.

The authors also are grateful to the nesting family of doves that kept them company outside the window as they worked, and to a strengthened friendship as a result of this book.

## Trademarks

All terms mentioned in this book that are known to be or are suspected of being trademarks or service marks have been appropriately capitalized. Alpha Books and Penguin Group (USA) Inc. cannot attest to the accuracy of this information. Use of a term in this book should not be regarded as affecting the validity of any trademark or service mark.

# Part 1

# So You're Going to Do a Science Fair Project

Welcome to the wonderful world of science fairs. If this is your first science fair experience, you're in for a fun learning experience. If you've already had some experience with fairs, you know what's involved and what to expect.

Chapters 1 through 6 lay out the science fair basics for those who are new to the game, and offer lots of advice and tips that even science fair veterans will find useful.

Be sure to read these chapters carefully before choosing one of the topics found in the later chapters. These chapters lay the groundwork that will assure smooth sailing once you begin your project.

# The Different Types of Science Fairs

## In This Chapter

- ◆ Understanding how a science fair works
- ◆ Deciding how difficult your project will be
- ◆ How will your project be judged?
- ◆ Taking your project past the local science fair
- ◆ Getting your project under way

Being required to complete one or more science fair projects has been a rite of passage—sometimes a welcome one and sometimes not—for generations of students.

If you're a student who enjoys science and science projects, good for you! You've discovered a fascinating and rewarding discipline that holds innumerable opportunities for careers, hobbies, and advancing the cause of humankind.

If you're not a science fan and find yourself staring straight-on at an impending science fair project, don't worry. You're not the only person in the world who is less than enthusiastic about science fairs. You've come to the right place for help with making a great project.

Before you even start thinking of an idea for a science project, let's take a few minutes to think about science fairs. How did they start? Are there fairs other than the ones held in schools? Do any of them give prizes or scholarships to winners?

In this chapter, we'll explore just what a science fair involves, how it's judged, and what opportunities there are to advance past your local fair. Let's get started.

# Just What Is a Science Fair, Anyway?

The standards and rules for science fairs vary from school to school, region to region, and state to state, depending on the level at which you're participating.

**Standard Procedure**

An online site called the National Student Research Center, founded in 1992 at the Mandeville Middle School in Mandeville, Louisiana, includes an electronic library and journals of research conducted by students. The site is not exclusively science based, but contains science-related information and materials. You can find it on the Internet at youth.net/nsrc/nsrc.html.

Basically, a science fair is a collection of projects intended to identify and solve scientific questions or problems. Some fairs only accept projects completed by individuals, while others allow projects by teams of students. Most science fairs require that entrants follow specific rules and regulations, such as observing space limitations and the types of materials that are permitted for use. Some schools require science fair projects from every student, while others give students the option to participate. The number of years in which a student may participate in science fairs also varies from school to school.

While some fairs accept projects dealing with only one category of science, such as chemistry or biology, most school science fairs allow students to enter projects from all branches of science.

Projects entered in most fairs are viewed and assessed by judges, who assign grades or rankings based on a set of clearly defined criteria. Science fairs can be complex and heavily regulated, or more casual and informal. The only way to know what your school requires is to make sure you get a clear set of guidelines from your teacher.

# How Hard Is This Going to Be?

You know how some kids are really, really into art or music or camping or baseball or field hockey? Believe it or not, there are kids all across the country who are equally passionate about science.

You've probably known some of these kids. They're the ones who dream of being astronauts and build make-believe space shuttles. They join their schools' science clubs, and they often ask for science-related gifts, such as chemistry sets, microscopes, or ant farms.

These kids, not surprisingly, are the ones who really get into doing science fair projects. In fact, some of them *live* to do science fair projects. That's because they're generally motivated, curious, and organized. They know that science is cool, and they look at it as a way to learn and discover.

Other kids fall more into a more ambivalent category. These are students who don't get particularly happy when science fair time rolls around, but neither do they run away screaming. Doing a science fair project for a kid in this group is sort of like going for a yearly checkup with the doctor. He doesn't look forward to it, but it's not a big enough deal to get bent out of shape over.

And then there is the "Please, no—I'll do anything but a science fair project!" group. These kids, as you've probably gathered, would rather stay home and clean their rooms on a Friday night than do a science fair project.

Sometimes this dread of undertaking a science fair project stems from a lack of interest in science in general, an inability to manage time, a lack of organization, or not understanding the directions and rules one must follow.

**CAUTION**

### Explosion Ahead

Choosing a topic that's too complex and difficult can make doing a science fair project tedious and unpleasant. On the other hand, choosing one that's too easy may result in boredom, and a feeling when you finish that you've wasted your time.

Regardless of where you fall on the science fair appreciation spectrum, there is an appropriate project for you. Chapter 2 tells you everything you need to know about selecting a project that you'll enjoy doing and learn from. For now, just remember that the beauty of a science fair project is that the topic you choose isn't an indicator of whether or not your project will be a good one.

While the most advanced, involved projects may get a lot of attention when they're viewed at the science fair, they're not necessarily the best projects. A simple, but well thought-out and executed project can be just as valuable as one that required months of work.

The trick is to choose a project that is appropriate for your ability, interest level, and the amount of time and money you'll have available.

Sure, some topics are more interesting than others, and you should always work to your ability level, but no science fair is judged solely on the complexity—or lack of the same—of a project.

In short, this means that if you work carefully, follow the guidelines of your school's fair, and use the scientific method (you'll learn all about that in Chapter 4), you can do a really good science fair project that's as simple or complex as is fitting for your skill level, interest, and time constraints.

Even if you're not excited about doing a science fair project, remember that the benefits exceed the science knowledge you'll be getting. In addition to technical skills, organizing and completing a science fair project can help you develop time management, organizational, and research skills. Preparing your exhibit involves creativity and graphic ability, and, if you need to make a presentation, your communication and public speaking skills will be called into play, as well.

# How Are Science Fairs Judged?

The manner in which science fairs are judged can vary, but most have similar point systems for evaluating projects, and most judges look for pretty much the same things.

Knowing and understanding what sorts of qualities judges will be looking for in your project can give you valuable direction during the course of your work.

Generally, science fair projects are judged according to the following standards:

- **Creativity.** When considering the creativity involved in a science fair project, judges will ask if the student exhibited a sense of curiosity when choosing a topic, if he or she was creative in designing the display, and if creativity was exhibited in the manner in which the student reached his or her results.

- **Scientific thought.** Judges will look to see if the student shows the use of the scientific method, including stating the problem, using appropriate materials, proper procedure, observation, and conclusion. They will look for original ideas having scientific value, consider the background investigation used, and look at how the student approached the project. They also will consider the analysis of results and the validity of the student's conclusions.

- **Thoroughness.** When considering the thoroughness of work involved, judges will look to see if all information presented is accurate, if sufficient data was collected, and if the project indicates that the student has a full understanding of his or her chosen topic. They also may consider whether there is sufficient documentation of the student's work.

- ◆ **Skill.** In this category, judges will look at the manner in which the project has been mounted or built. They'll consider the neatness and durability of the display, and ask if the project reflects the student's own work. They'll also keep an eye as to whether the project meets all safety standards and other fair guidelines.

- ◆ **Communication.** When considering communication, judges will ask if the project is self-explanatory and easily understandable to the average person. They'll look for correct spelling and grammar, and whether signs, lettering, and diagrams enhance the display or result in its being cluttered and confusing. They'll also ask whether the project is logical, clear, and complete.

**Standard Procedure**

Most students use computers for making signs, graphs, and so forth. If you don't have the use of a computer, you can make these things by hand, as long as they're done neatly and clearly.

Again, judging standards and criteria may vary among science fairs. You probably will receive a copy of your fair's judging standards when you submit your project topic. If you don't, ask what basis the project will be judged on, and make notes so you can refer to them while you're working.

# Different Levels of Competition

Most likely, you're thinking about or preparing for your school's science fair. School fairs are the most common, but not the only types of science fairs held. There also are regional, state, and national fairs.

## Local and Regional Science Fairs

Local fairs include school science fairs, but also could be sponsored by a community group or other organization. A local chapter of the Boys and Girls Club of America, or a local civic organization, for instance, may sponsor a science fair for its members or area youth. Other sponsors of local or regional fairs include newspapers, colleges and universities, industries, and scientific and engineering societies.

Regional science fairs normally include the best projects from local fairs. A regional event could be on a county level, or some other designated area that includes several school districts or other local science fairs.

**Standard Procedure**

For a fairly extensive list of regional, state, national, international, and virtual science fairs check out physics.usc. edu/~gould/ScienceFairs. The site is hosted by the Department of Physics and Astronomy at the University of Southern California.

Regional science fairs tend to be well organized and structured, and often offer awards donated by business, industry, or scientific organizations.

The Southern Appalachian Science and Engineering Fair, for instance, held in 2003 at the University of Tennessee in Knoxville, offered prizes including scholarships, cash, and savings bonds. It also offered special awards including an educational photo kit from the Eastman Kodak Company, and a medallion and certificate from the Herbert Hoover Presidential Library Association.

Winners of various categories of regional science fairs may be invited to participate in larger fairs. In the Southern Appalachian fair, for example, four senior projects were invited for inclusion in the Naval Nationals, a science fair sponsored by the Secretary of the Navy program in the Office of Naval Research.

## State Science Fairs

Many states also have statewide science fairs, which are open to students who excelled in the local and/or regional fairs. Many of these fairs include large numbers of exhibits, and offer significant rewards. They may be hosted by a college or university, a business or industry located in the state, or the state itself.

An example of a statewide fair is the California State Science Fair, hosted by the California Science Center and in its fifty-second year. Nearly 1,000 participants from throughout the state are expected to compete in 2003 for awards totaling more than $50,000.

New Mexico has held a state science fair since 1953. Called the New Mexico Science and Engineering Fair, it is sponsored by the state itself, and the New Mexico Institute of Mining and Technology. The fair receives financial support from those and other organizations within the state. The four finalists in the state fair are eligible to participate in the Intel International Science and Engineering Fair.

The sponsors of your local science fair should be able to tell you about regional and state fairs in which you may be able to participate.

## National and International Science Fairs

While national fairs generally are restricted to students who reside and attend schools in the country in which they're held, international fairs may include projects from all over the world.

The Intel International Science and Engineering Fair, held each May in a major city, includes more than 1,200 students in grades 9 through 12 from 40 different countries.

These students compete for more than $3 million in scholarships, internships, scientific field trips, and tuition grants. The grand-prize winner gets an all-expense-paid trip to attend the Nobel Prize ceremonies in Stockholm. The Intel fair was started in 1950 by Science Service, a nonprofit organization based in Washington, D.C. While many organizations support the science fair, Intel is its primary sponsor.

Entrants must have participated first in a local, regional, or state fair that is an affiliate of the Intel International Science and Engineering Fair.

Another Science Service–affiliated fair, this one for students in grades five through eight, is the Discovery Channel Young Scientist Challenge. Forty finalists compete every October in Washington, D.C.

Other national and international science fairs include the Siemens Westinghouse Science and Technology Competition, the Great Australian Science Show, the Belgian Science Fair, Canada Wide Science Fair, and Costa Rica's Feria Nacional de Ciencia y Tecnología. Also, Ecuador's Youth National Fair of Science, Technology and Innovation; the European Science Fair; Ireland's Esat BT Young Scientist and Technology Exhibition; and the New Zealand Science Fair.

## Virtual Science Fairs

It was only a matter of time until science fairs went virtual, and the time has come. Virtual science fairs—sometimes called "environmentally friendly science fairs"—are popping up on the Internet, and getting increasing amounts of attention.

With some big names like NASA sponsoring the online contests, some students are participating in virtual fairs in addition to traditional ones, or even substituting virtual fairs for the real thing.

The point of a virtual science fair is to challenge students to use information technology tools to discover and explore the future of science and engineering. Students are presented with real-life problems, and asked to solve them with technology.

All information is submitted and judged electronically.

# The First Steps to Getting Started

If you're going to do a science fair project, make up your mind right now to get organized, and to remain so throughout the project.

While keeping careful notes and taking care of equipment might seem like a hassle, you'll be very glad down the road that you did so.

A good step to take early on is to set up a timetable that will guide you through your project. Once you've decided on a topic and can know a little better what it will entail, try to figure out how much time you'll need for each step. If you're going to be sprouting bean seeds, for example, you'll need to allow at least a week for the experiment, and possibly two weeks. Remember that you'll also need time to write a report, assemble your display, and research your topic.

Chapter 2 gives you lots of suggestions for finding a topic that will be suitable for your ability level and result in an interesting project. For now, you can start thinking about general areas in which you're interested in working.

If you love animals, for instance, you may want to consider narrowing your focus to a project that deals with biology. If stars and planets are your thing, look to the field of astronomy for inspiration.

No matter what project you choose, however, staying organized and making sure that you complete each part of your work on time will help to assure you'll be successful.

## The Least You Need to Know

- Science fairs vary greatly, but basically they are collections of projects intended to identify and solve scientific questions or problems.

- A great science fair project does not need to be overly difficult, as long as it's done properly and follows the necessary guidelines and principles.

- Understanding how a science fair project is judged can help you be attuned to possible problems as you proceed with your work, and help you to generate a successful project.

- There are different levels of science fairs—from local to international.

- The very first steps to getting started include making a timetable, narrowing your topic field, and committing to getting and staying organized.

# Choosing Your Topic

## In This Chapter

- ◆ Choosing the all-important topic
- ◆ It's all about learning
- ◆ Asking questions to generate topic ideas
- ◆ Choosing a topic within your interests, costs, and guidelines
- ◆ Keeping your project fun
- ◆ Working within your abilities

Selecting a topic is often one of the most difficult aspects of a science fair project. You don't want to spend weeks or even months working on a project that you don't enjoy. You need to choose one that will be challenging enough to be interesting, but not so hard to do that you end up tearing your hair out.

You should choose a project from which you'll end up learning, and, ideally, it should be in an area in which you have a greater interest than one science fair.

In this chapter, you'll learn all about choosing a topic that will lend itself to a good science fair project. In order to do that, you have to consider a bunch of factors, and think about what works best for you. Your best friend may have had a great time growing crystals, but if you have no interest in how or why crystals form, that won't be a good project for you.

Always look for a project in which you have a sufficient amount of interest. If you don't, you won't be excited about doing it, and it will become a chore instead of an adventure.

# Understanding the Purpose of a Science Fair Project

You might think a science fair project is all about conducting an experiment, building a *backboard*, and getting a decent grade from your teacher. But it's not. A science fair project is all about learning—from the minute you start thinking about possible topics, until the time you've finished writing your final report.

> **Basic Elements**
>
> A **backboard,** in case you're wondering, is a freestanding display on which you mount your science fair data. The backboard, which displays the title of your project, along with charts, graphs, the abstract, hypothesis, and so forth, stands on a table.

The beauty of having students do science fair projects (and probably the reason that so many schools require them) is that they provide learning on many different levels.

Doing a complete science fair project not only requires science-related knowledge, but forces you to embark on many other endeavors. These include research, writing, displaying, and building. A science fair project also can be a great way for students to learn about being responsible for their time, and about being innovative, resourceful, and neat.

Some of the main areas of learning that occur during the course of a science fair project include research, writing, general science knowledge, organization, and presentation. Let's look at each.

- ◆ **Research.** Research is vitally important to a science fair project, and doing a project forces you to become familiar with basic research techniques. Chapter 3 covers this subject in detail. Learning to be a proficient researcher will be extremely valuable to you throughout your school career and beyond.

- ◆ **Writing.** People who can write comfortably and capably have an easier time in many situations than people who can't. It's no fun to have to spend hours agonizing over how to say what you want to say. Just like math or art, writing comes easier to some people than others. The more you do it, however, the easier it becomes, and the better at it you get. Writing a research report or paper as part of your science fair project is excellent practice for writing you'll need to do not only while you're in school, but throughout your life.

◆ **General science knowledge.** In order to even pick a topic for a science fair project, you've got to have a basic knowledge of the branches of science. In order to work through a project correctly, you've got to know about and use the scientific method. Learning about these general areas of science gives you a good base of knowledge that you can apply both now and in upper-level classes.

◆ **Organization.** While we're not saying that kids and teenagers should be organized to the point where they never misplace a house key or forget to write down a phone message for Dad, organization is absolutely necessary when conducting a scientific experiment, writing a report, or assembling a display. If you can't keep track of your materials, or at what point you are in the experiment, you'll risk doing the experiment incorrectly and invalidating your results. Organizational skills are important in many areas of life, and employing them well while doing a science fair experiment is a great way to develop them.

**Standard Procedure**

Try to think about other areas of life in which you'll use the skills necessary for completing a science fair project, such as organization and research. If you take a few minutes to think about it, chances are you'll come up with dozens.

◆ **Presentation.** You can have the greatest science fair project in the world, but if it's sloppy and unreadable, you're not going to be winning any blue ribbons. You don't need to be an artist to come up with a display that's neat and pleasing to look at. If you don't have a good eye for balance, color, and other "artsy" qualities, you can ask someone who does for some suggestions. You'll also learn more about presenting a project in Chapter 28. Getting acquainted with some of the guidelines that pertain to the aesthetic value of displays, however, will ultimately benefit you.

While these are some of the major learning areas included in a science fair project, they are by no means the only ones. Undertaking a science fair project will result in your gaining knowledge on many levels, and will serve as an excellent learning process. Now, let's get back to how to choose a great topic.

# Identifying a Problem

When choosing a topic for your project, the first thing you'll do is to make an observation, or identify a problem. A great way to do this is simply to watch carefully what goes on around you. If something happens that makes you wonder, chances are the question could be adapted into a science fair project.

If you've always been intrigued at the way your dog turns around three times before he finally settles in next to the TV for his nap, for instance, consider that as a project possibility. Do all dogs turn three times, or do some turn only once? Do any dogs turn five or six times before lying down?

All you'd have to do to turn that question into a science fair project is to identify it as a problem, then use the scientific method to work through the problem. Maybe you'd end up observing 16 different dogs and coming up with the average number of times they turned around before settling down to sleep.

> **Explosion Ahead**
>
> Regardless of what topic you choose, you've got to apply the scientific method as you put your project together. If you don't use it, your project won't be scientifically valid, and you may be disqualified from your science fair.

The world is an exciting, intriguing place, with millions of questions to be answered. If you train yourself to observe what's happening around you, you'll never run out of questions. And, as long as there are questions, or problems, you'll always be able to come up with science fair projects.

# Science Divided: Natural vs. Physical

All of science can be divided into two major fields: natural science and physical science. Natural sciences are those that deal with the study of all living things. Physical sciences deal mostly with the areas of matter, force, and energy.

Natural sciences, sometimes called biological sciences, include many areas, such as the following:

- ◆ Biology
- ◆ Botany
- ◆ Biochemistry
- ◆ Behavioral and social sciences
- ◆ Microbiology
- ◆ Medicine and health
- ◆ Gerontology
- ◆ Zoology
- ◆ Food sciences
- ◆ Physiology

> **Standard Procedure**
>
> To get a better idea of project categories and how they might overlap with other scientific areas, have a look at the categories for the 2003 California State Science Fair. They're listed online at www.usc.edu/CSSF/Info_Genl/Categories.

Physical sciences include areas such as these:

- Chemistry
- Physics
- Geology
- Astronomy
- Oceanography
- Computer science
- Engineering
- Mathematics
- Environmental science

The topic of your science fair project will fall under either natural or physical science, and likely under one of these categories. Each of those categories, however, could have many subcategories.

Figuring out which category your project belongs in may not be as easy as you would imagine. You might assume, for instance, that a project to determine the average number of times a dog turns around before resting would fall under the area of zoology. But it could be classified as behavioral science, which includes both people and animals.

Many areas of science overlap, which may make it difficult to decide which category your project belongs in. Take your best guess, and then check with your teacher or advisor to make sure you're on the right track.

# Relating Your Topic to Your Interests

Many teachers and judges say students don't do well with science fair projects because they choose topics in which they're not sufficiently interested.

If you don't like to be around animals, don't undertake a project in which you try to determine if male rats are stronger than female rats.

Consider your career goals (if you've formulated them), your hobbies, the activities in which you participate, and other special interests you may have. If you're looking forward to a career as a computer scientist, for instance, it might make sense to do a computer science–related project. If you're planning to be a doctor, a topic concerning a health issue would be a reasonable choice.

If you love sports, chances are you could come up with a sports-related topic, such as the one found in Chapter 11 or the project described in Chapter 21.

If you enjoy being outside, a biology, zoology, or earth and space science project would be a good choice. If shopping is your passion, try to come up with a project in which you test different types of fabric for strength or durability.

Get personal and think about what you like to do. You and your science fair project are going to be spending a lot of time together for the next few weeks. Make sure it's something you not only can live with, but can enjoy.

# Making Your Project Fun

If you choose a topic you're really interested in, chances are that you'll have a good time conducting the experiment and learning more about the subject.

Take a few minutes to think about what you like to do, and how you can choose a topic that would be really fun to turn into a science fair project.

If you like to make things blow up and explode, consider the project described in Chapter 22. If you fancy yourself to have some inventor's blood in you, you might enjoy the project described in Chapter 25.

Before settling on a project, take a little time to look at all the projects described in this book. They involve all levels of work and offer a variety of projects and opportunities to learn about different subjects. Take a few minutes to think about which ones sound like fun.

# But What If I Really Hate Science?

If you truly dislike science, and, just like with math or writing or art there are people who do, you can still come up with a doable science fair project.

A trick to try is to narrow down your topic until it no longer seems like science. It's good to have a narrow focus for any science fair project, but sometimes it's possible to extract a "nonscience" project from a larger idea.

Recall the mention a few paragraphs back of testing different fabrics for strength and durability. Now think about how you might do that in a nonscientific way. If you think about it, perhaps you'd come up with an idea such as securing four squares of different types of fabric to the bottom of your backpack for two weeks. That should—in a nonscience kind of way—give you a pretty good indication of which fabrics hold up to wear, and which ones don't.

Of course, you'd still have to make observations and follow the steps of the scientific method. But you wouldn't be working in what we think of as a traditional science setting, which might make the project a little more enjoyable.

# Keeping Your Project Within Your Ability Level

The downfall of many a science fair participant has been choosing a project that's either too easy or too difficult. If the project is too easy, it becomes boring. If the project is too difficult, it becomes overwhelming.

## Knowing If It's Too Easy

If you can run through the experiment in your head before you've even started exploring the idea you have for a project, chances are that it's going to be too easy.

If you've already done a project, it's fine to build on it and develop it into something more advanced, but repeating a project because you think it will be less work isn't a good idea.

This book divides projects into three categories: beginning, intermediate, and advanced. None of the projects in this book are overly complicated nor are the experiments overly difficult. If you're wondering if you should choose a beginner's project or an intermediate one, go for the more advanced topic.

Be honest with yourself when you're assessing a topic idea. You'll probably know whether or not it's going to be too easy. If it is, pass it up for a harder one. Spending time on a project that is not at all challenging really is a waste of time.

## Knowing If It's Too Hard

A project may be too difficult if it involves too many areas of research, and your research skills aren't good enough to keep up. If you have to research three or four topics to have the information you need for your project, you might want to consider changing topics.

Of course, if you've got crackerjack research skills and like nothing better than discovering new information, then go ahead and tackle the more difficult project.

Another indication that a project may be too difficult is if the experiment makes no sense to you. When you study the experiment, it should appear to be logical. If you're coming up with the experiment on your own, you should be confident that you've properly identified the control and variable, that you can formulate a reasonable hypothesis, and that you fully understand the procedure. Be sure that you have a dependable method of recording data, as well.

If you're unsure whether a project is too difficult, read through it several times and try to get a good idea of what it entails. It's always good to push yourself a little, but if the project is overly difficult, you'll end up frustrated, and chances are that the judges will not be impressed.

## Finding a Topic That's Just Right

A topic that's at your ability level is likely to be centered around a subject you've studied in the not-too-distant past, or will be studying soon.

**Standard Procedure** _____

Basing your science fair project on a topic you've recently studied or will be studying in school is a good idea because it not only assures the project will be within your ability level, but that you'll have a basis on which to build your project idea. You'll be able to relate information you discover as you do your project to what you've learned—or will be learning—in the classroom.

You should feel comfortable when you read about the procedure, have an understanding of the type of materials involved, and feel that you can manage the project—not that you're in over your head.

Trust your comfort level. Don't be afraid to take on a project that's challenging, but if you're really uncomfortable with one that seems way too difficult, don't hesitate to drop down a level.

Once you've chosen a topic, try to stick with it, unless it becomes impossible. Researching your topic and getting organized for the experiment are time-consuming and sometimes difficult tasks. Starting over could set you back tremendously and jeopardize your chances of finishing the project on time.

# Considering the Cost of Your Project

There are many science fair projects—many of which are presented in this book—that require only readily available materials. You probably can find most of the materials you'll need for the majority of the projects in this book around your house.

Some projects, however, require special equipment, and that equipment could cost you some money.

If you decide to do an astronomy project that requires a telescope, and you don't own or can't borrow a telescope, you're going to have to spend significant dollars to get one. If you want to learn whether male rats are stronger than female rats, you'll have to buy rats with which to work. If your project requires you to use a grow light to raise plants, you may need to purchase one.

Before you commit to doing a specific project, make sure you know what materials you'll need—and how much they'll cost. If you don't want to shell out money for materials, there are many projects you can do that will cost you very little, or nothing.

# Keeping Science Fair Guidelines in Mind

Many science fairs have more than their share of rules, and it's extremely important that you know the rules before you nail down a project topic.

Many fairs, for instance, do not allow living animals to be displayed, although you may be able to work with them in an experiment and record your findings in charts, graphs, or photographs.

The Intel Science Talent Search, a science fair held every year for the most promising high school science students in America, has an extensive list of rules and regulations to which participants must adhere. A sampling of those rules includes:

- Common laboratory animals must be legally acquired from reputable animal breeders.

- Alcohol, acid rain, insecticide, herbicide, and heavy metal toxicity studies are prohibited.

- For the purposes of student research, all body fluids, including saliva and urine (but excluding hair), are to be considered tissues.

**Standard Procedure**

Remember that if you're not allowed to display plants, animals, or other materials at the science fair, you can still work with them for your experiment. Use photographs and other tools to document and record your results.

- ◆ Microorganisms collected, isolated, and/or cultured from any environment should be considered potentially pathogenic.

- ◆ Students under 21 years of age may not purchase and/or handle smokeless powder or black powder for science projects.

- ◆ Informed consent is strongly recommended for all projects using human subjects, and is required for all subjects when more than minimal risk is determined.

Rules that you'd be likely to encounter for a middle school, junior high, or high school include:

- ◆ No living creatures, including plants and microbes.

- ◆ No chemicals, including caustics and acids.

- ◆ No projects with unshielded belts, pulleys, chains, or moving parts with tension or pinch points.

- ◆ No human or animal parts except teeth, hair, nails, animal bones, histological sections, and wet-mount tissue slides.

Make sure you read *all* the rules and regulations before choosing a topic, and don't try to get around them or take shortcuts. You don't want to work hard on a project, only to have it be disqualified because you're in violation of a regulation.

## The Least You Need to Know

- ◆ The main purpose of a science fair project is for you to learn and increase your skills in various disciplines.

- ◆ Coming up with an idea for a science fair project is as easy as asking a question about something you observe, then using the scientific method to attempt to solve the problem.

- ◆ All areas of science can be divided into natural science or physical science.

- ◆ Relating your topic to something you're interested in and would have fun exploring will assure that you'll remain interested and involved in the project.

- ◆ Choosing a topic that's too easy can result in a loss of interest, while one that's too difficult can result in frustration and confusion.

- ◆ You should consider fair regulations and costs involved in a project before choosing your topic.

# Chapter 3

# Learning What You Can About Your Topic

## In This Chapter

- ◆ Saving time and effort through research
- ◆ Using the library to obtain information
- ◆ The advantages and possible problems of Internet sources
- ◆ Finding an expert to help you
- ◆ Giving proper credit to the sources you use
- ◆ Staying organized during research for best results

Before you undertake a science fair project, it's a good idea to do some advance research and learn what you can about the topic you've chosen.

You may be required to complete a research report or research paper as part of your project. Some fairs require this, while others do not. If you do have to write a research report, you can read exactly how to do that in Chapter 30. Reading Chapter 3 first, however, will give you a lot of valuable information about conducting research that you'll need even before you get around to writing your report.

Even if you don't have to submit a research paper or report, it's important to understand the basics of researching and gathering information. Knowing these will help you as you proceed through your project. If you're going to experiment with plants, for instance, it's a good idea to get to know a little bit about the types of plants you'll be using. This gives you a broader base of knowledge and will help you to understand and appreciate what occurs during the course of your project.

By learning what you can about your topic in advance, you'll be able to be more organized and aware during the time you're working on your project. In this chapter, you'll learn a lot about finding and organizing information. And you'll learn about documenting your research.

# Keeping Yourself and Your Project Organized

We live in a world in which there sometimes seems to be too much information. We get information via e-mail, television, and radio. Information comes through newspapers, magazines, and an entire World Wide Web.

For many people, the problem with researching is not being able to locate adequate material—it's coming up with a system to keep the material organized, verifying its reliability, and knowing how to use it effectively.

If you anticipate gathering a great deal of research material, think about how you might organize it. Here are some suggestions:

- Use different colored folders to organize material. You might use red, for instance, to hold papers with information you'll need to write your research report. A green folder could hold miscellaneous information, a yellow folder your list of sources, a blue folder your handwritten notes, and so forth. Do whatever works best for you.

- Use a three-ring binder from which you can add and remove papers as you wish. This allows you to keep your file from becoming overly large and unmanageable. Just be sure to save any papers you remove in a folder in case you need them later.

- If you're keeping research notes and are comfortable using a computer, consider setting up an electronic file. You can keep any notes you have on paper in a folder, but you might

---

**Scientific Surprise**

A worldwide survey conducted by Reuters in 1996 revealed that two-thirds of all people in management positions were suffering from increased tension and one-third from ill health due to information overload. A psychologist named David Lewis analyzed the survey results and came up with the phrase "information fatigue syndrome" to describe the symptoms the managers experienced.

transfer them onto your computer. Many people find it easier to manage documents electronically than shuffling through papers.

You probably have had enough experience with schoolwork and projects to know which method makes the most sense for you. But remember that staying organized will make your project go a lot more smoothly and be easier for you in the long run.

# How to Avoid Reinventing the Wheel

There probably is not a science fair project you can think of that hasn't already been done—or at least one very similar to it. And while you by no means want to copy someone else's work, neither do you want to start at square one if you don't have to.

Conducting some advance research can go a long way in helping you to avoid unnecessary problems and work. Let's say that you need three healthy plants for your project. Through your research, you learn that the plants you're working with dislike bright light. Having that knowledge will, hopefully, prevent you from exposing the plants to too much light and jeopardizing their health. Doing so will save you the trouble of having to grow or buy more plants, as you'd have to do if your original ones died.

You would have used information that someone before you learned and shared, thereby preventing you from making a time-consuming and perhaps costly mistake.

# Where to Go for Information

School research used to be primarily conducted in libraries, and much of it still is. The techniques used to find information may be different than they were 10 or 15 years ago, but the library still is the place to go for much of the type of research you'll need to do for a science fair project.

The other obvious research site is the Internet. If you have a computer at home from which you can access the Internet, you'll have at your fingertips a great deal of information. Again, the trick may not be finding enough information, but keeping it organized and knowing how to verify and use it.

A third source, and one you might not as readily think of, is a person who is well acquainted with your topic and who would be willing to

> **Scientific Surprise**
>
> The original concept of an information web was introduced by Vanneyar Bush in 1945 in an article in the *Atlantic Monthly*. In it, Bush described a photo-electrical-mechanical device called a Memex, which could make and follow links between documents on microfiche.

work with you. We'll explore that possibility a little later in this chapter. For now, let's step into the library.

## The Library

Libraries in general have changed dramatically during the past 10 or 15 years. Most libraries have replaced their old-fashioned card catalog systems with computerized systems. Many local libraries are hooked into a larger library system, such as a county system. This gives the smaller libraries the advantage of being able to borrow books from larger libraries—usually within a couple of days of the time you request them.

> **Standard Procedure**
>
> Generally, the best thing about libraries is the people who work in them. Librarians tend to be extremely helpful, and will go out of their way to see that you have what you need. Don't hesitate to ask for help if you need it.

The only materials you may not be able to have sent from the main library in your area to your local library are reference books, which generally are not permitted to leave the building. This means you may end up having to make a trip to the main library to access materials you'll need.

When using reference materials in the library—either your school library or the public facility—be sure to record all the necessary source information before you leave. Many researchers have been tripped up by not immediately recording sources of information.

Let's say that over the course of two or three days of working, you found some really great, important information that will dramatically help you with your project. You took notes on what you learned, but you forgot to write down where the information came from.

> **Basic Elements**
>
> A **bibliography** is a list of all the sources from which you obtained information for your research paper. It can include books, online sources, magazines, government publications, newspapers, journals, and reference materials such as encyclopedias.

When it comes time to credit sources and write a *bibliography* for your research report, you'll have no idea of which information came from which books, and you probably won't even be able to remember which books you used. (We'll get into citing sources and documenting information later in this chapter in the section called "Keeping Your Sources Straight," and in more detail in Chapter 30.)

While libraries once were primarily stiff, silent places, many of them have loosened up considerably, now offering all sorts of fun activities for kids and programming for adults. And most libraries have computers from which you can access the Internet.

This is a great benefit if you don't have Internet access at home. Don't forget to ask the librarian for help if you need it.

Remember that libraries are home to all sorts of nifty periodicals that you might find useful. Look for publications such as *Nature, Science, Popular Science, Discover Magazine, Smithsonian, National Geographic,* and *Science News.*

## The Internet

While most people think the Internet will never fully replace libraries, there's no question that it's become an incredibly important research resource. The ability to access information from around the world via a personal computer has changed the way millions of people work, think, and live.

*The Internet is an important resource for your science project research.*

While you're probably already familiar with how to use the Internet, there are some specific topics to think about when you're using it for research purposes. They include the following:

♦ Knowing how to best utilize search engines, such as Google or AltaVista, and subject sites, such as Yahoo!, to assure you get adequate information about your topic.

♦ Being able to find some specific, useful sites on your own.

♦ Having the skills to verify the information you find on the Internet.

A *search engine*, in case you're not sure, is a tool that enables you to gather information from across the Internet by typing in some keywords pertaining to your research topic. The engine searches databases of web pages and spits out a list of sites that it thinks will provide the kind of information you're looking for.

Anyone who has ever used a search engine, however, knows that this method is not perfect. You're likely to end up with a lot of information, but it could be of no value to you.

Some people prefer *subject sites*—or Internet subject directories—to search engines. A subject site, such as Yahoo!, is a collection of websites that have been gathered and organized by humans—not computers. You begin by selecting a very general subject, such as "Science." Next, you choose the topic you're interested in—say "Biology"—from a big list of science-related areas.

Next—you guessed it—you again choose a topic (let's say "ecology") from another list. And so on and so forth, until you narrow it down to your science fair topic, which happens to be on the effects of acid rain.

When using either a search engine or a subject site, it's important to have a good handle on what you're looking for. If you're using a search engine to get information about how acid rain affects crop yields, for instance, you should not simply use "acid" and "rain" as the keywords you type into the search box.

**Explosion Ahead**

Some big search engines have logged keywords from more than 1 billion online documents. If you type in a common word, you run the risk of being inundated with websites that may, or may not, contain useful information. This can result in a lot of confusing information, and a big waste of time on your part.

**Basic Elements**

A **search engine** is an electronic procedure that locates web pages that have been identified in advance based on keywords. A **subject site** (sometimes called an Internet subject directory) is a collection of websites gathered and organized by humans according to subject matter.

You'll be swamped with websites, the great majority of which won't contain the kind of information you need. If you type in the keywords "acid," "rain," "affect," "crop," "yields," however, you'll get a much narrower—and more useful—listing of applicable sites.

Some popular, well-rated search engines include the following:

- Google
- HotBot
- AltaVista
- Lycos
- Infoseek
- Excite

Some useful subject sites include these:

- Yahoo!
- WWW Virtual Library
- Subject Guides A–Z
- Britannica.com
- Librarians' Index to the Internet

To find these subject sites, type the name of the site into the keyword box of one of the search engines.

While search engines and subject sites are useful, they're not always necessary. You may be able to identify applicable websites on your own and go directly to them. For instance, if you're looking at a helpful magazine in the library, chances are you can find a website for the publication listed somewhere within its pages. Most magazines have sites that contain at least some of the information found in the paper copy.

Government websites also are excellent sources for scientific information, and can be accessed directly. To find practically any government site available, go to www.SearchGov.com. You'll get a complete list of links to take you to government agencies such as the Department of Agriculture, the Energy Department, and so forth. Another helpful tool is the Federal Web Locator, found at www.infoctr.edu/fwl.

**Standard Procedure**

It's a good idea to print out Internet pages from which you've obtained information. Not only will you be able to recheck the information and verify sources, the printed page most likely will include the site's web address, allowing you to revisit it if necessary. It's also a good idea to keep a written record of all Internet sites from which you obtain information.

A vast amount of information is available on the Internet, enabling you to learn about countless topics. While much of this information is completely verifiable and reliable, some of it is misleading—or just plain wrong.

It is a good idea to stick to reputable websites, such as those of colleges and universities, the government, and well-known, respected organizations such as the American Psychological Association or American Medical Association.

Regardless of what website you use, you should use the following criteria to evaluate the information it contains:

◆ **Legitimacy.** Is the organization or person sponsoring the site a legitimate, responsible group or person? Does the site include contact information other than an e-mail address? Would you be able to contact the organization or person by phone or mail if you wanted to? Does the site include a statement saying that the information it contains has the official approval of the organization?

**Standard Procedure**

The Internet Public Library, a site hosted by the University of Michigan School of Information, has a great page in its teen section called "Learning to research on the Web." You can access this site at ipl.si.umich. edu/div/teen/aplus/internet. htm. It's got a lot of good information that you'll find helpful.

◆ **Accuracy.** Are facts backed up by sources so they can be verified? Is the information presented clearly, without typographical and spelling errors? Is it clear who is responsible for the information contained on the site?

◆ **Bias.** Does the material presented include opinions or information that appear to be one-sided? Are opinions stated as such, or presented as fact? Does the information include advertising?

◆ **Timeliness.** Is the information up-to-date? Does it tell you when the content was last updated? Are press releases or other news pieces current?

Once you've become experienced with using the Internet for research purposes, you'll begin to be able to recognize which sites are good and which ones aren't. For now, try to stay with sites hosted by agencies or groups that you've heard of and know are legitimate.

## People Who Can Help

Written or online information is great, but don't overlook the human touch. Maybe your mom's best friend is a science teacher who could recommend some great books or steer you to other sources of information.

Maybe your uncle knows somebody who is retired from a science-related job and would be happy to talk to you about your project. Or, you could call the science department of an area college or university and ask if there's a professor or upper-level student who might be willing to give you some help.

Something to remember is that most people enjoy helping others—especially young people. Be polite and sincere when you ask for help, and you're likely to be pleasantly surprised by the responses you get.

# Giving Credit Where Credit Is Due

When you're writing a research paper, or even just researching the topic of your science fair project, it's extremely important that you acknowledge the sources of your information.

You cannot—ever—take information directly from any source and use it as your own without giving credit to that source.

This rule applies to material you find in books and periodicals, sources such as government publications, and the Internet. If you take information from a source and use it on a paper or in a report, you must give credit where credit is due.

According to the Writing Tutorial Services at Indiana University in Bloomington, credit must be given for information used in all of the following circumstances:

♦ When you use another person's idea, opinion, or theory

♦ When you use any facts, statistics, graphs, drawings, or other pieces of information in any form that are not *common knowledge*

♦ When you quote another person's spoken or written words

♦ When you paraphrase (put into your own words) another person's spoken or written words

As you can see, that about covers it. When in doubt, give credit to your source. And be sure to record the source of your information as soon as you find it. Don't think you'll remember in which book or on which website you read something.

**Basic Elements**

**Common knowledge** is facts that can be obtained from a wide variety of sources, and is recognized as true by many people. Common knowledge does not need to be documented. If you wrote in your report, for instance, "Albert Einstein was a great scientist," you would not have to document that statement because it is common knowledge.

Not crediting the sources of your information can result in huge problems, so don't even think about trying to get away with it.

## Plagiarism Is Plagiarism, Whether You Mean to or Not

If you use information without giving credit to your source, it's called *plagiarism*. Plagiarism is a serious matter, and teachers, professors, and advisors are increasingly on the lookout for it. There are even mechanisms now that scan websites to detect plagiarism.

**Basic Elements**

*Plagiarism* is the act of taking ideas, facts, quotations, or other information from any source and claiming it as your own by not giving credit to the source from which it came.

To put it simply, plagiarism is stealing. It's taking something that doesn't belong to you, and claiming it as your own. It's a form of theft, which, as you know, is a very serious matter.

Even if you don't intend to steal somebody's idea or information, if you copy something that is not common knowledge and don't give credit, that's exactly what you've done.

Let's say, for instance, that you're doing a report on Albert Einstein. It's fine for you to say in your paper that Albert Einstein was a great scientist (because that's common knowledge). It's not okay, however, for you to write without citing your source that in 1944 Einstein contributed to the war effort by copying by hand the paper on special relativity he had written in 1905 and auctioning it off for $6 million.

That information, by the way, was taken from an article written by J. J. O'Connor and E. F. Robertson. It was found on an Internet site posted by the School of Mathematics and Statistics at the University of St. Andrews in Scotland.

To not credit that source would have been plagiarism. To avoid plagiarizing, remember to follow these guidelines:

◆ Write the information you use in your own words. That doesn't mean just changing a couple of words or the order of facts or figures. A good way to do this is to read the information you plan to use, think about it, and then say it in your own words to a family member or friend. Many people find it easier and more natural to talk than to write. After you've said it, just write down what you said.

◆ If you copy information word for word from a source when you're taking notes, be sure to put quotation marks around it. That way you'll know those words came from the source with which you were working—not from your own head—and you'll be able to give credit to the source.

◆ Once you've written information in your own words, it's a good idea to check your writing against the original. It's easy to use particular words or phrases that you've read, thinking that they're your own.

◆ Avoid the temptation to claim someone's opinion about a specific topic as your own, even if it's an opinion with which you agree. It's fine to write, "I agree wholeheartedly with Einstein's opinion concerning the atomic bomb." But it's not all right to rewrite his opinion and pretend you thought of it on your own.

> **CAUTION**
>
> **Explosion Ahead**
>
> Plagiarism is considered a serious problem on college campuses, and is taken as a very serious offense. On many campuses, a student found guilty of plagiarism risks being expelled from the school.

Recording and crediting sources can be time consuming and tedious, and your teachers may or may not require that you do so at this stage of your school career. Be assured, however, that as you get older, you will need to understand and follow the guidelines pertaining to crediting sources and plagiarism.

## Keeping Your Sources Straight

Researching a topic or writing a research paper or report can result in the gathering of a lot of information. That's why it's really important to stay organized and keep track of what information came from which sources.

Sitting with pages and pages of notes in front of you and no idea where the information came from is not an enviable position. You'll end up doing the bulk of your work twice if you need to go back and find out which sources you took the information from.

Consider these suggestions to make sure you keep your sources straight:

◆ Take careful notes as you read information you think you can use, noting the name of the book, periodical, agency, website, or other source; the author; the date of publication; the name of the publisher, if applicable; the website name, if applicable; and the location of the publisher, if applicable.

◆ Use a copy machine to photocopy large portions of material instead of trying to write it by hand.

◆ Use a special notebook or note cards for your research, and keep them in a safe place. Losing the information means you'd have to start over.

◆ Don't rely on your memory when researching—ever. It's tempting to think you'll remember where you found a particular piece of information, but chances are you won't, especially if you use a number of sources.

Researching a topic can be hard work, but it also can be interesting and rewarding. Learning and understanding the basic methods of research will make the process easier. Once you've written your first two or three papers, you'll be working like a pro.

## The Least You Need to Know

◆ For many people, the problem is not finding enough information while researching, it's keeping the information organized and verifying that it's credible.

◆ Researching your topic before you actually begin your science fair project can help you avoid making costly and time-consuming mistakes.

◆ The library, the Internet, and people who know a lot about your topic are the best sources for getting information you need.

◆ It is very important to keep track of where you find information you'll be using in a report or research paper.

◆ Copying someone else's written work is called plagiarism, for which the consequences can be very serious. Your project most likely would be disqualified from the science fair, and you'd risk a failing grade.

◆ Preparing yourself to be organized when gathering information will save you a lot of time, trouble, and wasted effort.

# Chapter 4

# Understanding and Using the Scientific Method

## In This Chapter

- ◆ Defining the scientific method
- ◆ The origins of the scientific method
- ◆ The importance of a clearly stated problem or question
- ◆ Researching your problem to reach a hypothesis
- ◆ Making sure you have everything you need

Three different people can witness the same event, and each come up with a different account of how it occurred. Police and investigators know that this is true; so do parents and teachers. You've probably had some experience with this phenomenon, as well. Have you ever had two friends who attended the same event give you completely different reports of what happened?

These conflicting viewpoints occur because we all see the world differently. We all have beliefs, biases, and perceptions that cause us to view things the way we do.

While these differences are what make us each unique and assure that the world is an interesting place, they can make it difficult to determine what is really true and what isn't.

Scientists over the centuries found they faced the same problems when it came to sorting out the truth from nontruths. To solve the problem, they devised a methodical framework within which to work. This framework is called the *scientific method*, and it's extremely important to your science fair project.

# The Scientific Method Made Easy

The scientific method is a tool that helps scientists—and the rest of us—solve problems and determine answers to questions in a logical format. It provides step-by-step, general directions to help us work through problems.

### Basic Elements

The **scientific method** is a series of steps that serve as guidelines for scientific endeavors. It's a tool used to help solve problems and answer questions in an objective manner.

You probably use the scientific method in everyday life without even realizing it. Let's say that one night you feel like reading in bed, but your mom has already told you three times that it's late and you need to keep your light out and go to sleep.

Because you know you're not going to be able to sleep regardless of what your mom says, you reach under the bed for your handy flashlight and flip the switch to turn it on. Nothing happens.

Now you're faced with the problem of not being able to read because your flashlight doesn't work, and you're not happy about it. Having identified the problem, you think back to the last time your flashlight didn't work, and you remember that it was because of worn-out batteries.

You guess that worn-out batteries are the reason your flashlight isn't working now, so you get some new batteries from the drawer next to your bed and replace the ones in your flashlight. Presto! Your flashlight works.

### Standard Procedure

There are two forms of the scientific method, but they both require the same objective reasoning and steps. The experimental method employs numerical data and graphs, while the descriptive method gathers information through visual observation and interviewing. The experimental method generally is used in physical sciences, and the descriptive method in zoology and anthropology.

Without realizing it, you've just worked through the steps of the scientific method to solve a problem.

For our purposes, there are five steps to the scientific method. They are …

◆ Identify a problem.

◆ Research the problem.

◆ Formulate a hypothesis.

◆ Conduct an experiment.

◆ Reach a conclusion.

When your flashlight wouldn't turn on, you knew you had a problem. That took care of the first step. Your research (the second step) was conducted when you thought back to the last time your flashlight didn't work and remembered that you needed new batteries. You completed the third step by coming up with the hypothesis (an educated guess) that you needed new batteries this time, as well.

You conducted your experiment (the fourth step) when you replaced the batteries and turned the flashlight on. When the flashlight worked, you reached the conclusion that indeed, it had needed new batteries. You completed the fifth step of the scientific method and proved your hypothesis to be correct. You also got to finish that great book you were reading!

So you see, the scientific method is not mysterious or difficult, although you can use it to work through some difficult problems.

# Don't Even Think About Not Using It

There's no question that you must use the scientific method when you do your science fair project. Nearly every science fair has rules that clearly state that the scientific method must be observed and followed. Not following the method's steps could cause you to arrive at incorrect results, and could make your entire project invalid.

Using the scientific method assures that you'll work objectively, not subjectively. That means that you won't bring to the project your personal, preconceived thoughts about what you think should or should not happen, or your own interpretations of your observations and data. The scientific method assures that you'll stick to "just the facts." You'll employ objective reasoning instead of subjective reasoning. The difference, when it comes to science projects, is extremely significant.

What is objective reasoning, and how does it compare to its subjective cousin? Simply, objective reasoning is when you recognize that there's a problem, then use research and experimentation to solve it. Sort of like when your flashlight didn't work. You thought about (researched in your head) the problem, then experimented by replacing the batteries. And, ultimately, you solved the problem.

Had you reacted subjectively, however, you would have recognized the problem of the flashlight not working, but you may have come to an invalid conclusion based on emotions or some bias instead of figuring out that you needed new batteries. Maybe you would have assumed that the flashlight wasn't working because it was red, and you always hated the color red.

Suffice it to say that when you're working on a science fair project, or dealing with any scientific problem, the scientific method, which encourages objective reasoning, is the only way to go.

# Whoever Thought of the Scientific Method, Anyway?

The scientific method probably is the cumulative result of hundreds of years of scientific pondering and people working to make sense out of what goes on around us.

> **Scientific Surprise**
>
> Francesco Redi was a man of many talents. In addition to his scientific work, he worked as a doctor, and was also a poet and writer. As a doctor, he stressed the importance of a balanced diet.

A scientist often credited with being the first to employ the scientific method—although he wouldn't have called it that—is Francesco Redi, an Italian physician who lived from 1626 until 1697.

Redi's work to disprove spontaneous generation is often credited as the first modern experiment, and was conducted within the parameters of what we now call the scientific method. Basically, Redi disagreed with the then-popular notion that some species appeared spontaneously in nonliving matter.

As odd as it seems to us, people believed for hundreds of years that certain species could be grown from nonliving materials. Some people, for example, thought that if you put worn, sweaty underwear in an open jar with some husks of wheat and let it sit for a few weeks, the sweat from the undies would penetrate the husks of wheat and turn the wheat kernels into (are you ready?) mice. Weird, huh?

Redi's work, however, didn't involve underwear and mice, but rotting meat and maggots. It was widely believed in those days that maggots appeared spontaneously from rotting meat. Redi, however was convinced from his research, which consisted primarily of observing activity surrounding rotting meat, that the maggots resulted from the eggs that visiting flies deposited on the meat.

He hypothesized that the maggots came from the eggs of flies, and set out to prove that he was right. Redi set up an experiment in which some meat was left uncovered, some was partially covered, and some was sealed so that nothing could get to it.

As you probably guessed, no maggots appeared on the sealed meat, because no flies could reach it. Redi's hypothesis was shown to be correct, and the roots for the scientific method had taken hold.

# Stating the Problem

The first step when using the scientific method is to state the problem you'll be attempting to solve. This step is sometimes referred to as "stating your purpose." You're identifying the purpose of the project, which is to solve a problem or answer a particular question.

In a science fair project, you deliberately identify and state the problem you're attempting to solve. In everyday life, you probably do the same thing in a more informal way. Did you ever ask yourself (or someone else) why your neighbor's grass is really green, and yours is sort of brown with lots of weeds in it?

Or why you got an 88 percent in the last math test, and your best friend only got a 74 percent? Every time you ask a question of this sort, you're stating a problem. You can probably go ahead and solve these types of problems by doing some informal research, making a guess, and checking to see if you're right.

You may learn, for instance, that your neighbor has been fertilizing his lawn, and your best friend forgot to take his math book home the night before the test.

**Standard Procedure**

Try to be aware of how you may identify problems as they occur in everyday life. Chances are that you'll find yourself using a simpler version of the scientific method to solve problems and answer questions on a daily basis.

When you state your problem or question, be sure not to make it so broad that it becomes unmanageable. Try to focus in and make your problem specific. This will help you to find a starting point in solving it.

For instance, stating your problem as "In what conditions do plants grow best?" is so broad and general that it would be almost impossible to know how to begin working through a project.

But, if you ask, "Do bean plants grow better in direct sunlight, indirect sunlight, or shade?" you've narrowed down your problem to address only one type of plant, and one factor affecting its growth.

Also, be sure that your problem is one that can be solved through experimentation. Solving the bean plant problem stated above can easily be accomplished through a controlled experiment. A controlled experiment is when you test a variable against a control. You'll learn all about controls and variables in Chapter 5.

# Researching Your Topic

Once you've stated your problem, you'll probably need to do a bit of research before you formulate your hypothesis. In Chapter 3, you learned about researching the topic that you were interested in pursuing.

At this point, you'll be refining your research to specifically address the problem you've stated. This will allow you to put forth an intelligent and well thought out hypothesis, which is simply an educated guess about the results of your project.

Remember to document your research, as discussed in Chapter 3, and use a variety of sources. Don't forget that your previous experiences and knowledge you already have can be valuable additions to your research.

# Coming Up with a Hypothesis

### Basic Elements

A **hypothesis** is an educated guess about the outcome of your experiment, based on knowledge that you have and research you've conducted.

The third step of the scientific method is formulating a *hypothesis*. All this means is that you'll need to come up with a statement concerning the predicted results of your experiment. It's what you think will happen, based on the research you've done and your knowledge. A hypothesis doesn't include *why* you think you'll get certain results, just what you think they will be.

The more you know about your problem, the better equipped you'll be to come up with a logical hypothesis.

## Giving It Your Best Guess

Your hypothesis should be clearly and simply stated, and should be in statement form—not a question. If you're guessing about the growth of the bean plants, for instance, the following statement is an example of a clear, concise hypothesis:

Bean plants will grow better in direct sunlight than in indirect sunlight or shade.

Because it's understood that a hypothesis is an educated guess, you don't need to say that you're guessing. You needn't say, for example, "I think that bean plants will grow better in direct sunlight than in indirect sunlight or shade."

## Remembering That It's Only a Guess

So what happens if you state your hypothesis, only to find out after the experiment that it's wrong? Nothing.

Guessing incorrectly the results of your experiment doesn't make the experiment wrong, or any less valuable than if your hypothesis turned out to be correct. Your hypothesis isn't necessarily the answer to your problem. It's simply a statement of what you think will happen.

### Explosion Ahead

If your hypothesis turns out to be incorrect, resist the temptation to change it. This defeats the purpose of using the scientific method. An incorrect hypothesis will not affect the quality of your project.

## Testing Your Hypothesis

The experiment that you conduct will be to test your hypothesis. Your experiment will be designed around your hypothesis, and will either prove or disprove it.

It will be important to conduct your experiment with your hypothesis in mind. However, it's imperative that you don't engineer your experiment to prove that your hypothesis is correct. You can't for instance, add Miracle-Gro to the direct-sunlight bean plants to assure that they'll grow better than the other plants. If you do, you'll be helping along your hypothesis, but invalidating the results of your experiment.

# Gathering the Materials You'll Need

As with any project, it's important that you have all the materials you'll need to conduct it properly. Just as you wouldn't begin making a recipe without first making sure you have all the ingredients, you shouldn't begin your experiment without making sure you have what you need.

Of particular importance will be having the proper tools for measuring. Many of the projects in this book require careful, accurate measuring. Most tools for measuring aren't fancy or expensive, but they're essential for conducting an experiment properly. Important measuring tools include tape measures, metric rulers or meter sticks, measuring spoons and cups, thermometers, and clocks or watches with second hands.

**CAUTION**

**Explosion Ahead**

Using materials you already have is convenient and economical, but just be sure that it's also safe. If your experiment calls for pouring boiling water into a glass beaker, for instance, don't substitute a lightweight, plastic cup, just because you happen to have one in your home.

## Using What You Have

Most of the projects explained in this book don't require much—if any—special equipment. This is intentional, because it's important that everyone has the opportunity to participate in science fairs without the need to purchase expensive scientific supplies.

Most of the projects require items that are commonly found in homes, such as glasses, plastic cups, paper towels, and so forth. Just be sure to check with whoever is in charge at your house before you take anything to use for your experiment.

## Knowing Where to Get Them

Some of the projects explained in upcoming chapters do require some equipment that you may not have access to at home. Most of what you need, however, can be easily purchased in a home supply store, a drugstore, or an electronics store such as Radio Shack.

If your experiment calls for a specialty item, or if you need a specialty science product for another reason, check out the list of scientific supply companies in Appendix C.

## The Least You Need to Know

- The scientific method is the result of hundreds of years of cumulative effort and learning, but Francesco Redi is sometimes credited with being the first scientist to use the method.

- Stating your problem clearly and concisely is the first step in using the scientific method properly.

- Keep your research focused around your problem, and use it to formulate your hypothesis.

- A correct or incorrect hypothesis doesn't affect the outcome of your experiment or the value of your science fair project.

- Make sure you have all necessary materials ready before you begin your experiment.

# The Meat and Potatoes of the Scientific Method

## In This Chapter

- ◆ Preparing for your experiment
- ◆ Keeping it simple
- ◆ Understanding controls and variables
- ◆ Writing down directions and keeping them close
- ◆ Working step-by-step
- ◆ Making sure you get it right

As you've no doubt realized, there are many steps to doing a science fair project. You've got to select a topic, which may require some research, organize your project, research some more, come up with a hypothesis, gather materials, and so on.

While all of these steps are important pieces of a science fair project, and should be conducted carefully and thoroughly, perhaps the most important part of your project is the experiment itself.

In this chapter, you'll learn all about performing the experiment that will either prove or disprove your hypothesis.

# Conducting the All-Important Experiment

The experiment is the nuts and bolts of your science fair project. It tests your hypothesis, and solves the problem you stated earlier in the project. Without the experiment, you'd have no way of knowing if your hypothesis is correct or incorrect, and you'd have no answer to your problem.

If you're going to do one of the experiments explained in this book, then your experiment is already designed, and you just need to follow the instructions. If you didn't have a set of instructions, however, you'd need to design your own experiment. In other words, you'd need to figure out a procedure through which you could solve the problem you had stated. You'd need to determine:

 ◆ What your controls and variables will be.

 ◆ Where your experiment will be conducted.

 ◆ What supplies and materials you'll need for the experiment.

 ◆ How long the experiment will take.

 ◆ What procedures you'll use during the course of the experiment.

**Explosion Ahead**

Try to avoid overly complicated experiments if you're a science fair novice. Attempting to conduct an experiment that's too complex puts you at risk for becoming overwhelmed and frustrated.

If you do design your own experiment, be sure to think the entire process through carefully. Remember to write everything down.

A good experiment depends not on the complexity of the subject matter, or the complexity of the experiment itself, but on a well-thought-out plan of action that will either prove or disprove your hypothesis. In fact, simple experiments often are best.

An example of a simple (at least by scientific terms) experiment that has become renowned is scientist Robert Millikan's oil drop experiment, conducted in the early part of the twentieth century. Millikan, who was a physicist at the University of Chicago, figured out the size of a charge on a single electron by measuring how tiny oil drops—sprayed out of a perfume atomizer—fell in his apparatus.

Keeping your experiment simple assures that you'll be able to conduct it in a logical, sequential manner and without problems. As you get older and more experienced, your projects will become more complex and difficult. Simple projects can earn blue ribbons at science fairs, however.

Work carefully and thoroughly when conducting your experiment, and don't forget to ask for help if you need it. Be sure to have everything you need for the experiment ready before you begin, and try to stay organized throughout the procedure.

If you want to, you can use the following checklist to make sure you'll be organized and ready for your experiment.

❏ I have a clear idea of the problem I'm attempting to solve through my experiment.

❏ I have researched the topic in order to develop a good hypothesis.

❏ I have a thorough understanding of my hypothesis.

❏ I know how the experiment will be conducted.

❏ I have a written, step-by-step procedure for my experiment.

❏ I know where I will conduct the experiment.

❏ I have allowed ample time for the experiment.

❏ I have all the necessary materials and supplies.

❏ I have an adult available to help me if necessary.

❏ I have the necessary tools to record my observations.

❏ I have considered how I will represent my experiment on the display board.

Once you fully understand the experiment you'll be doing and have completed the steps on the checklist above, you'll be well equipped to begin an extremely important part of your science fair project.

# Using Controls and Variables

Science is different from other disciplines in that it's, well … scientific. One hundred people can read the same piece of poetry and have 100 different ideas about what it means. A political science class can argue until the cows come home about whether Bill Clinton was a good president or a bad president—and never reach a consensus. An auditorium full of music lovers can listen to the same concerto and walk away with many different reactions.

The scientific method, and in particular the controlled experiment, sets science apart from these other disciplines. A controlled experiment results in a conclusion. Fifty people who witness the same experiment will see the same results. But, for this to occur, the experiment must be done properly, with the proper controls.

When you conduct the scientific experiment as part of your science fair project, you'll be working with controls and variables. Understanding what controls and variables are and the roles they play in your experiment is very important.

## What Are Controls and What Do They Do?

Basically, a *control* is something that remains the same, or constant, throughout your experiment. Controls are sometimes called *controlled variables*, or constant variables.

A variable is one thing in an experiment—a single factor—that is changed and, because it is changed, will affect the outcome of the experiment. There can be more than one variable, but each one must be isolated and tested separately so you can observe the effects of each one. Variables can be changes intentionally caused by the person conducting the experiment, or responses that occur as a result of the actions of the experimenter.

Changes caused by the person conducting the experiment are called *independent variables*. An experiment should have only one independent variable.

Changes that occur as the result of an independent variable are called *dependent variables*. If you're confused, consider this example.

**Basic Elements**

A **control**, or **controlled variable**, is a factor that remains the same throughout an experiment. An **independent variable** is a change introduced by the person performing the experiment. A **dependent variable** is a change that results from the independent variable.

Let's say that you're trying to figure out if bean plants that are watered with distilled water grow faster than those watered with good old tap water.

The independent variable is the type of water that you will use to water certain bean plants. You, the person conducting the experiment, are causing the variable to occur. Changes that occur to the bean plants that are watered with each type of water are the dependent variables. The changes to note would be how well each plant grew as a result of being given different types of water.

The controlled variables, or controls, in this experiment would be that the plants receive the same amounts of the different types of water, are the same type and size, are kept in the same environment, and so forth. The controls must assure that everything about the plants is the same, except for the type of water they're given.

The best way to see if an experiment will work is to define your variables. Variables are anything that can affect the outcome of your experiment.

*These pots represent a control and a variable in a scientific experiment.*

## Examples of Controls and Variables

Controls and variables can be a little confusing, there's no question about it. Take a minute to think about these examples of controls and variables. They may make the whole issue easier to understand.

- Optimum conditions for a plant to grow include correct amounts of light, water, and nutrients. If your experiment is growing plants, such optimum conditions would be your control. If you change the conditions in any way, you've created a variable. You would only test one of these possible variables at a time, so you know which variable affected the plant growth. If you introduce more than one variable at a time, you won't know which one caused the difference in growth.

- Chapter 22 contains an experiment aimed at determining whether brand-name sandwich bags are better than store brands, also called generic brands. In that experiment, the generic-brand bags are the control group, and the various name-brand bags are the variables. One at a time, the experiment tests the variable against the control group.

- The experiment in Chapter 7 uses food kept under normal circumstances as the control group, and the same foods, placed in sealed plastic bags and distributed to different areas of a home, as the variables.

- In Chapter 14, the task is to discover whether pennies minted before 1970 are heavier than those made later. A bunch of pennies made before 1970 is the control group, while pennies minted after 1970 serve as the variables. Each group of pennies made after 1970 is compared to the same group of pre-1970 coins.

◆ If you wanted to test laundry detergents, your control would be a dirty shirt washed in plain water. Independent variables would be different types of laundry detergent in which shirts containing the same type and amounts of dirt are washed. You would need to be sure that the shirts were all washed in water that comes from the same source, and is the same temperature in order to be able to tell which detergent worked best.

If you have questions about controls and variables, study these examples to better understand. Controls and variables are a very important factor in the scientific method.

# Taking It One Step at a Time

Following the steps of the scientific method is much like following a recipe.

Let's say that it's Mother's Day, and you wake up feeling like you should do something special for Mom. After all, she's hauled you around to all your practices, sat on the hard bleachers in the rain to watch your soccer game, and usually even has something ready for you to eat when you get home.

Because you know your mom has a serious sweet tooth, you decide to make her a cake. Chocolate, of course. And you want the cake to be really good.

Even if you're an experienced baker, you wouldn't begin making a cake by throwing a bunch of stuff in a bowl, but by carefully reading a recipe to see what materials you would need and what procedure you would follow.

## Writing Down the Directions

Likewise, if you were designing your own experiment, it would be very important to write down the procedure you planned to use and have it in front of you when conducting the experiment.

Not only would this assure you'd remember all the steps and be able to proceed as you'd planned, but it also would make it possible for you, or others, to repeat the experiment at a later time.

If you design an experiment and write instructions, make sure they're clear and easy to follow. Not only do you need the explanation of how your experiment is conducted, judges will want to read through it and be able to visualize how the experiment proceeded, as well.

**Standard Procedure** _____

If you design and write the instructions for your own experiment, it would be a good idea to have a teacher or another student read through the directions before you conduct it. It's sometimes easier for another person to spot a potential problem, such as steps that are out of sequence or materials you may have forgotten to list. You might receive some good suggestions and ideas.

## Following the Directions

Regardless of whether you write the directions for your experiment yourself, or use those printed in this book, it's important that you follow them carefully. Work in order of the steps listed, and have everything ready that you'll need as you proceed, just like when baking that cake for Mom.

If you don't follow the directions and proceed in the order of steps as written, you could end up with inconclusive results or an invalid experiment.

# Repeating the Process to Assure Accuracy

It's always a good idea to do an experiment more than once.

Repeating your experiment provides trials to assure there are like results. That is, if you perform the same experiment in the same manner, the outcome should be the same. If you do, you'll know that the results are accurate. If you don't, you'll need to try to figure out what happened.

Perhaps you inadvertently introduced an extra variable to the experiment. There are many factors that can change the outcome of an experiment, and that's why it's extremely important to keep everything as uniform and controlled as possible.

If you get different results in different trials of your experiment, think about possible variables. Remember the following:

◆ A lot can happen if you're working with plants. Watch out for drafts, different temperatures that could cause different rates of plant growth, differences in watering or feeding, and so forth.

◆ If you're using a heat source or a cooling source, such as a fan, it needs to be a consistent temperature in all trials.

◆ Foods being tested must be maintained at the same temperatures and humidity levels.

◆ If you're dissolving substances, you must be sure the liquid in which they're being dissolved is the same temperature. Temperature affects the rate at which a substance will dissolve.

◆ If your experiment involves people or animals, such as in Chapter 12, be aware that factors such as age, agility, and overall health could be variables.

◆ When using specialized equipment or materials, be sure you have exactly what you need, and use the same equipment for every trial. Substituting something could cause different results.

**Explosion Ahead**

If you're doing an experiment that involves timing, don't be tempted to think that mistiming by a second or two won't matter. Science experiments must be performed as accurately and precisely as possible. A few seconds doesn't seem like much, but it could make a huge difference in your results.

These are just a few suggestions of potential variables that could affect the outcome of your experiment. If you conduct the experiment several times and get the same result, however, you can be assured that your results are valid.

Also, it's sometimes necessary to conduct an experiment more than one time so that you can average the results. This is especially true if the experiment involves timing, such as in Chapter 23.

Remember that if you'll need to repeat an experiment that takes significant time to perform, you'll need to allow an adequate period in which to complete the work.

When planning for and conducting a scientific experiment, it's extremely important to be organized and to complete your work in as controlled a manner as possible. By doing so, you'll be assured that your results are valid and your project is a valuable one.

## The Least You Need to Know

◆ Simple experiments are as valuable as complicated ones, and help assure you'll be able to complete them successfully.

◆ It's important to understand that there are several types of variables, including a controlled variable, which is often simply referred to as a control.

◆ Having written directions in front of you and following them step by step, as written, is crucial to conducting an accurate experiment.

◆ It's often necessary to repeat an experiment several times to assure sound results.

# Chapter 6

# Collecting Data and Making Observations

## In This Chapter

◆ Knowing the proper way to take measurements

◆ Measuring more than once to assure accuracy

◆ Understanding the metric and English systems

◆ Observing a science experiment

◆ Using charts and graphs to present your data

◆ Conducting analysis and drawing conclusions

Science fair projects provide wonderful learning opportunities, and can be a great deal of fun to do, as well.

A good project, however, does require careful and accurate work. While some students—all right, many students—enjoy planning projects and conducting experiments, they seem to balk at certain aspects of a science fair project. These generally include tasks such as writing reports, collecting data, and making careful notes regarding their observations.

This type of work may be a bit tedious for some students, and lacks the glamour of an exploding lunch bag (see Chapter 22) or a homemade lava lamp (see Chapter 25). It is extremely important, though, that you pay close attention to the data that relates to your project, and your observations of what occurs during your experiment.

These things must be carefully recorded in order to be of use to you in the later steps of your project.

# Taking Measurements and Recording Them Correctly

Nearly all experimental science projects include measurements of one sort or another. You may have to measure the height of some plants you've grown, or how far a balloon rocket flies along a string.

You may need to measure and record time, calculate the amount of liquid left in a container after evaporation has occurred, or measure ingredients necessary for an experiment.

Because measurement is an integral part of many science fair projects, it's extremely important that you use proper measuring techniques and tools, and understand different types of measurements such as length, volume, and temperature.

## Making Sure You're Doing It Right

Regardless of what type of measuring you're doing, you must work carefully and accurately. And, you must make sure that you use the measuring tools—such as scales, tape measures, and beakers—correctly.

Here are some suggestions to help you make sure you measure accurately during the course of your science fair project:

♦ When measuring liquid in a beaker, cup or graduated cylinder, hold the container at eye level and read the volume of the liquid at the lowest point of the *meniscus.*

**Basic Elements**

A **meniscus** is the curved edge of a liquid in a container such as a graduated cylinder. To measure accurately, you must read the quantity from the bottom of the meniscus.

However, if you're measuring a liquid that has an inverted meniscus—such as liquid mercury—you'll need to read the measurement at the top of the curve.

♦ If using a bathroom scale to weigh an object, be sure the scale is calibrated to zero. That means that it registers exactly at zero before an object is placed on the scale.

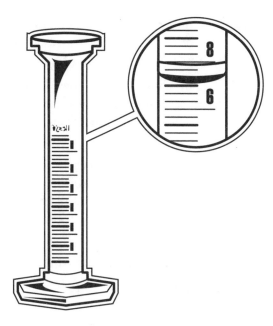

*A graduated cylinder.*

♦ When measuring length, make sure that the edge of your measuring stick or tape is right up against the edge of the object you're measuring. If using a measuring string, be sure that it's stretched tightly along the object being measured.

♦ When measuring temperature, be sure you hold the thermometer at eye level when reading it. Read the number inside the glass column of the thermometer at which the red liquid stops.

♦ Repeat your measurements whenever possible to assure accuracy.

Many inaccuracies result from improper measuring techniques. Be sure you always work carefully to assure your work is precise.

**CAUTION**

### Explosion Ahead

If you're measuring the temperature of water or another liquid, don't remove the thermometer from the liquid in order to read it. Leave it in the liquid and read it from within the container. Exposing it to the air will change the reading.

## Understanding Units of Measurement

Americans use a system of measurement called the English system. This is the system you're probably familiar with. You know—gallons, pounds, yards, and so forth.

American scientists, however, along with scientists from all over the world, use the metric system. While you may not want to use this system because you're not overly familiar with it, you should know that most of the world has used the metric system, also known as the International System of Units, for some time.

The United States, however, continues to use the English system of measurement, which causes trade problems and raises issues with other countries.

The metric system is based on the decimal system. You multiply by 10 each time you move to the next larger unit, and you divide by 10 each time you move to the next smaller unit. The metric system's multiples and submultiples are always related to powers of 10.

Once you understand the metric system, it actually is a lot easier to use than the English system, which is based on a multitude of units of measurement.

In the early 1960s, a revised system of units of measurements called the International System was adopted. The revised system is based on seven prescribed base units. Those units are as follows:

- The meter (m) to measure length

- The gram (g) or kilogram (kg) to measure mass

- The second (s) to measure time

- The ampere (a) to measure electric current

- The Kelvin (K) to measure thermodynamic temperature

- The candela (cd) to measure light

- The mole (mol) to measure quantity of matter—it's commonly used in chemistry to refer to an amount of a compound or element

In addition to the units of measurement of the International System, a metric unit commonly used to measure volume is a liter (l).

Measurements in this book are presented in both the English system and the metric system so you can get an idea of how the two compare. It's a good idea to understand the metric system so that you're not wary about using it if it becomes necessary. There have been proposals within the United States for years to switch to the metric system and get in tune with the rest of the world.

So far, however, for whatever reason, the United States has resisted changing. As you read through the chapters that describe the science fair projects, try to pay close attention to the metric measurements that are presented along with the English ones.

# What's Your Take on What's Going On?

The observations you make during the course of your science fair experiment are extremely important. Equally important is how you keep track of those observations.

The only really safe way to assure that your observations will be valuable is to record them as they occur. Don't think that you'll remember in two weeks something you observed today. You need to write down everything you see, using language that is clear and accurate.

## Don't Forget That the Little Things Count

Science is science, and must be conducted accurately and methodically. If you're making daily observations over a two-week period, conduct the observations at the same time every day.

When measuring plant growth, for instance, you'll need to measure the plant every 24 hours to get a daily reading. You can't measure a plant at 8 P.M. one day, 7 A.M. the next, and 3 P.M. the third day.

Pick a time that's convenient for you to conduct your observation, and stick with it.

When you make your observations, write down as many details as possible. If your plant has only grown a quarter of an inch in the past 24 hours, but has developed yellow spots on two leaves as well, be sure to record both of those developments.

**Standard Procedure**

A great idea is to start a journal at the very beginning of your science fair project and use it to record all of your thoughts and observations. It also can be useful for keeping track of data. The journal doesn't have to be fancy, just something you'll keep available and refer to when necessary.

Look for and record even the tiniest changes in color, temperature, and so forth. Maybe you'll see a tiny bit of rust beginning to form on iron nails, or seeds that are just beginning to germinate. Be observant, and remember to record everything you see. If you can, it's a good idea to take photographs to help you record what's occurring with your project.

## Keeping Your Observations Objective

When you're making observations, it can sometimes be tempting to "see" something that's really not there, or to ignore something that is present or occurring.

**CAUTION**

**Explosion Ahead**

Don't assume that the plant watered with orange soda (or whatever) is going to be the largest, even if it's inches above all the other plants halfway through your experiment. Continue observing and recording what you see until the very end of the experiment.

This could happen because you have a set of expectations regarding the project, or you're hoping that your hypothesis is correct.

If you guessed in your hypothesis that the plant watered with orange soda would be the least likely to survive, for instance, you might lean toward thinking it looks unhealthy, even if it's actually flourishing.

It's vitally important to avoid these tendencies. Although they're understandable, you must be completely objective when making observations. You must note and record exactly what you see, not what you would like to see.

Your job is only to observe and record what you see. At the end of the experiment, you can interpret your data and reach conclusions based on all your data and observations.

# Using Charts and Graphs to Present Data

Once you've finished making observations and collected all your data, you'll probably want to transfer the information you've recorded in your journal to a chart or graph. This will help you to clearly and concisely present your information.

There are three basic types of graphs. They are:

♦ **Bar graph.** A bar graph is used to show relationships between groups, such as different liquids and plant growth. A bar graph is good for showing large differences. Bars on a bar graph can run either vertically or horizontally, but the more common method is to use vertical bars. A bar graph uses an x axis and a y axis to represent two different types of data.

♦ **Line graph.** A line graph is used to plot ongoing data. A line graph also has an x axis and a y axis.

♦ **Pie graph.** A pie graph, or circle graph, shows the relationship of a part to the whole. A pie graph works very well for explaining percentages.

**Standard Procedure**

When using a bar graph or line graph, make sure it's clear what the x and y axes represent. Remember that the x axis is horizontal, and the y axis is vertical.

When presenting information on a graph, be sure to include an appropriate title and a key. If you're using a bar graph, for example, showing the growth rate of three different kinds of plants, you should use a different color for each plant.

Graphs are great, but charts can be equally valuable for recording information, especially if there are many parts to the information.

Also, you can use a chart to represent data from a graph in an organized manner.

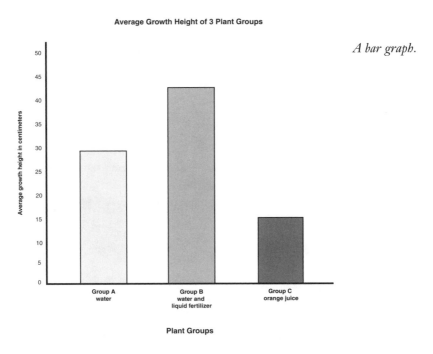

*A bar graph.*

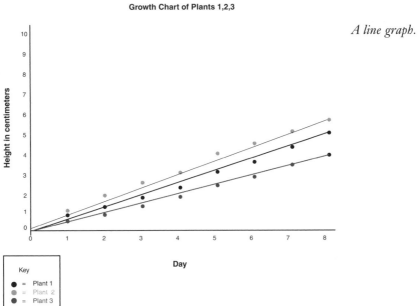

*A line graph.*

*A pie graph.*

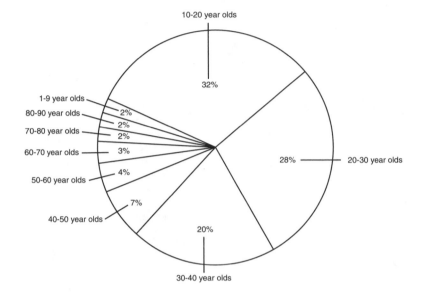

**Percent of people in age groups who scored 10 or more points with Bop it extreme**

10-20 year olds — 32%

20-30 year olds — 28%

30-40 year olds — 20%

40-50 year olds — 7%

50-60 year olds — 4%

60-70 year olds — 3%

70-80 year olds — 2%

80-90 year olds — 2%

1-9 year olds — 2%

*A chart.*

**Height of Plants measured in centimeters**

| Day | Plant 1 grown with just water | Plant 2 grown with water and fertilizer | Plant 3 grown with orange juice |
|---|---|---|---|
| 1 | 0.8 | 1.1 | 0.6 |
| 2 | 1.3 | 1.9 | 0.9 |
| 3 | 1.8 | 2.6 | 1.4 |
| 4 | 2.4 | 3.2 | 1.9 |
| 5 | 3.1 | 4.3 | 2.4 |
| 6 | 3.7 | 4.7 | 2.8 |
| 7 | 4.6 | 5.3 | 3.5 |
| 8 | 5.2 | 5.9 | 4.1 |
| 9 | 5.7 | 6.6 | 4.8 |
| 10 | 6.6 | 7.5 | 5.5 |

No matter how you choose to present your data, make sure that it's neatly done, easily readable, and clearly presented.

Your chart or graph is representative of all the work you've done on your science fair project. It's what the judges will see, and will affect the grade you receive.

A final bit of advice concerning charts and graphs is to keep them as simple as possible. Students sometimes are tempted to make them overly complicated in hopes of impressing the judges. The risk, however, is that you'll end up confusing the judges, which won't be beneficial to your grade.

**Standard Procedure**

You can create charts and graphs on a spreadsheet program on your computer. Some programs to consider are Excel or Microsoft Works. These programs will produce graphs from data you type into the program.

# Analysis and Conclusion

Once your data has been collected and presented by using a chart, graph, or concise and organized report, you'll need to logically and objectively interpret it.

Be sure that you first complete any calculations that will help you to draw a conclusion, and that all applicable results have been tabulated.

It's very important that you are as objective as possible when interpreting data. You must work only with the results that came from your experiment.

## Were You Right or Wrong?

When you've presented your data, analyzed it, and drawn conclusions, it will be clear whether your hypothesis was accurate or not. If your hypothesis was correct, great. If it wasn't, it takes nothing away from the value of your experiment.

Many scientists have been surprised at what they've discovered during the course of an experiment. If you discover something completely different than what you thought you would, don't worry—that's fine.

Once you've determined whether your hypothesis was right or wrong, you should write a short, concise summary telling how your results related to your experiment. This summary often is just one sentence, summing up what occurred during the experiment and how it relates to the results.

Your summary might also include your thoughts about why you got the results your did, or about what occurred during your experiment. Some teachers require a report that includes your thoughts and observations, while others ask only for a summary statement.

## The Least You Need to Know

- Nearly all experimental science projects include measurements of one sort or another.

- Each type of measurement must be done properly and accurately in order to make sure it will be correct.

- Measurements should be taken more than once to assure accuracy.

- While the American public uses the English system of measuring, our scientists, along with nearly everyone else in the world, use the metric system.

- It's important to pay attention to every detail while making observations during a science fair experiment.

- Record all of your observations as they occur, rather than thinking you'll remember what you've seen and recording them later.

- Graphs and charts are excellent means of presenting data in a clear and logical manner.

- It's extremely important to be objective and impartial when conducting analysis and drawing conclusions concerning your science fair project.

# Part 2

# Science Projects for Beginners

Your first science fair project can seem a little overwhelming. It might be the first major project you've ever done.

If you're feeling a little concerned about having to do a project, relax. Once you choose one of the experiments laid out in Chapters 7 through 11, you just need to follow the step-by-step directions that each chapter contains.

Take a few minutes to look over all the topics presented in these chapters before choosing the one you want to do. Remember to stick with what seems interesting to you. And remember to have fun with the project! Science isn't meant to be dull and all serious.

# Chapter 7

# Which Foods Do Molds Love Best (and Other Great Biology Projects)?

## In This Chapter

- ◆ Learning about mold
- ◆ Practical reasons to study molds and fungi
- ◆ Predicting your results
- ◆ Conducting your experiment
- ◆ Taking your findings to the next level
- ◆ More biology project ideas

You run into the kitchen after a hard morning on the soccer field, and you're so hungry you can't wait to eat. You grab the bread from the bread drawer and the cheese from the fridge. You get some juice and some potato chips, and you're ready to make a yummy cheese sandwich—your favorite!

But when you pull out a couple of slices of bread, you find that there's something green and disgusting growing on them. And then you notice that the cheese is covered with little white spots. Yuk! What the heck is it?

It's mold. It not only looks gross on foods, some kinds of it can make you sick if you eat it. Where did it come from? How did it get on your bread? Is it growing on any other foods in your house? It wasn't on the chips. Or in your juice, either. But what about those cookies you stashed under your bed a week or two ago? Or the beef jerky your brother keeps in his desk drawer? Will those foods be moldy when you go to retrieve them? Yikes!

The facts are that molds grow better on some foods than others, and that various factors contribute to their growth. You'll be learning a lot about molds in this chapter. Are you ready to get started?

# So What Seems to Be the Problem?

In this project, you'll be trying to figure out on what kinds of foods, and under what conditions, molds grow best. Working in an orderly, organized manner, you'll conduct an experiment that will help you to solve the problem.

You'll learn a lot about molds as you work through this project. You'll find out, for instance, if they prefer light or darkness, wet foods or dry foods, and heat or cold. You may be able to figure out when you've finished why there was mold on the bread and cheese you planned to use for your sandwich, but not on the potato chips.

**Basic Elements**

According to the U.S. Department of Agriculture, molds are microscopic fungi that live on plant or animal matter. Unlike bacteria, which are one-celled, molds are made of many cells. When viewed under a microscope, molds look like skinny mushrooms.

If you want to, you can use the title of this chapter, "Which Foods Do Molds Love Best?" as the title of your science fair project. Or you can think up your own title, or use one of these:

- ♦ Comparing the Growth Conditions of Molds
- ♦ Where and How Do Molds Grow Best?

Once you've got a clear idea of the questions you'll be trying to answer, keep reading to learn a little bit more about mold, and to think about the point of this science fair project.

# What's the Point?

Why do we care about how mold grows? Or care about mold at all, for that matter? Wouldn't it be better to just ignore this mold business and hope it goes away?

For starters, scientists think that molds have been around for about three billion years, making it highly doubtful that they'll disappear anytime soon.

Second, the study of molds has led to much knowledge and many benefits, including the discovery of penicillin, a medicine obtained from a mold called *Penicillium notatum*. Penicillin was the first antibiotic drug used, and is credited with saving countless lives. Other beneficial molds are those used to age and flavor cheeses (think blue cheese), and those used to make soy sauce.

If you like to drink soda, you need to give credit to the molds that manufacture the citric acid used to flavor many soft drinks. And certain types of fungi are used to improve soil conditions for farming.

The study of molds also has revealed which ones are dangerous and should be avoided. Some molds can cause allergic and respiratory problems, while others produce toxins that can be extremely harmful—even in small amounts.

> **Scientific Surprise**
>
> Penicillin, the original antibiotic, was discovered in 1929 by Dr. Alexander Fleming. Fleming learned that penicillin had the ability to stop the growth of a colony of germs placed into the same petri dish. Other doctors and scientists expanded on Fleming's discovery and penicillin eventually became known as "the miracle drug."

You've probably already encountered quite a few molds and fungi, some of which can be quite surprising. Listed below are a few facts you may not have known.

- If you have, or ever have had, athlete's foot, you can blame a fungus for your misery.

- The mold that grows on bread is called black mold, even though it doesn't always appear black.

- If you ever see a white fungus growing on the tropical fish in your aquarium, watch out. That fungus can be harmful, and sometimes even fatal to fish.

- Did you know that the mushrooms on your Friday night pizzas are actually a type of fungus?

- Warm, damp conditions in your bathroom make it the perfect breeding spot for a mold we commonly refer to as mildew.

- Slime mold—wouldn't you love to get a look at *that?*—is a type of mold that's primarily found in damp, moist woodlands. It grows on decaying logs and leaves.

- Certain large trees and ornamental plants like roses can become infected with a variety of fungus diseases, leaving the plants helpless to overcome the blight.

- A terrible fungus called *Phytophthora infestans* struck and destroyed Ireland's potato crop in the years between 1846 and 1850, causing more than one million people to starve to death and millions more to flee the country. That period became known as the Irish Potato Famine.

As you can see, there are many practical reasons to learn about molds and fungi. They have affected people, animals, and plants for thousands of years, and will continue to do so.

Knowing how molds grow on and affect the foods you eat can help you protect that food, and prevent or slow down the damage caused by molds.

In this science fair project, you'll work with five samples each of three different foods: bread, cheese, and oranges.

**CAUTION**

**Explosion Ahead**

Molds and fungi produce and release tiny spores, some of which can cause allergic reactions, asthma, and other problems for humans. For that reason, it's not a good idea to allow mold to be exposed within your home, and that's why you'll enclose the foods on which you want to grow mold in sealed bags.

Your control group (if you need a refresher on controls and variables, see Chapter 5) is those foods, as they are normally found in your house. For instance, the cheese control would be the cheese that's normally kept in your refrigerator. The bread would be the loaf stored in the plastic bag it comes packed in, and the oranges would be those that are kept in your house in the fridge or in a basket or bowl on the counter.

The variable group you'll work with is the same foods, but they will be sealed in zipper-lock bags and placed in different areas of your home. You'll need to create different environments in which to store the foods in order to conduct your experiment.

# What Do You Think Will Happen?

Every home has areas with different temperatures and humidity levels. Your bathroom, for instance, is a humid spot. If your home has an attached garage, it's probably cooler than inside your house. The inside of a dresser drawer that's rarely opened has a different *environment* than the kitchen. The windowsill on which the sun shines every morning is much different from the damp, dark cellar.

In this experiment, you'll be locating five different growing environments within your own house. Look for locations that are dry and dark, cold and dark, moist and warm, brightly lit, dark and damp, and so forth.

Once you've located the different growing environments, you'll be placing three bags in each of the five locations. One bag will contain oranges, one bread, and the other cheese. You'll need to take daily observations of the food in all of the bags.

What you need to do now, before you start the experiment, is to come up with a hypothesis, or a guess about what will happen. Do you think that the foods you keep in a dresser drawer will get as moldy as those stashed in the bathroom?

**Basic Elements** _____

Every living thing has an **environment**. The environment is all the conditions, circumstances, and influences that surround and affect the development of an organism. Environment is important to every living creature—from molds to humans.

**Standard Procedure** _____

Be sure to check with an adult before you start putting food samples in the bathroom or a bureau drawer. And be sure to place the bags where they'll be out of reach of pets and little brothers and sisters.

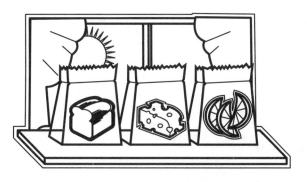

*Locate different environments within your home to determine where mold grows best.*

Think about where you may have observed molds and fungi in your home. Are there particular areas where you've noticed mildew or a moldy smell? If so, consider that when you're making your hypothesis. And think about what foods you may have seen molds growing on in the past. What similarities were there in those foods? Were they all kept in the refrigerator? Were they foods that had been sitting around the kitchen for a while because nobody liked them, or food that had already been cooked?

Some foods are more hospitable to molds than others, and molds grow better in some environments more than others. In your experiment, you'll be working to determine which foods molds grow best on, and in which environments.

Consider all the information you may already know. Then you can make an educated guess about the results of your experiment.

# Materials You'll Need for This Project

The materials you'll need for this project are minimal and easy to get. They are as follows:

- ◆ Five slices of bread (You can use any type of bread you have available, but all five slices should be the same.)

- ◆ Five slices of orange

- ◆ Five slices of cheese

- ◆ Fifteen zipper-lock sandwich bags

- ◆ One permanent marker

You don't necessarily need to put a whole piece of bread or an entire piece of cheese into each bag. A half piece will do just fine. And you should be able to get five slices out of a single orange.

# Conducting Your Experiment

Follow these steps to conduct your experiment:

1. Using the permanent marker, label five bags "bread."

2. Label five bags "orange slice."

3. Label five bags "cheese."

4. Place a piece of bread into each of the appropriately marked bags.

5. Place a piece of orange into each of the appropriately marked bags.

6. Place a piece of cheese into each of the appropriately marked bags.

7. Securely close and seal each bag.

8. Select one bag containing bread, another containing an orange slice, and another containing cheese. Group these three bags together as SET 1.

9. Repeat Step 8 to make four more sets of each food.

10. Place each set of three bags in a location of your home that has a different growing environment.

11. Using the permanent marker, label each bag with its location. Each set of three bags will have the same location marked on the bag.

12. Observe each bag once a day for two weeks. Try to look at the bags at about the same time each day in order to allow fairly equal growth time. Record everything you notice about the contents of each of the 15 bags on a chart like the one in the next section, "Keeping Track of Your Experiment."

**Standard Procedure**

Marking each bag with its proper location assures that you'll be able to return a bag to the right place in the event that your dog gets hold of it and proudly delivers it to your bedroom. Or when your little brother grabs a bag off the kitchen counter and stashes it in your mother's pocketbook.

13. At the end of two weeks, place all of the sealed lunch bags in the kitchen trash bag. Mom won't be at all happy with you if you leave bags of moldy food lying around.

It's very important that you make daily observations during the course of your experiment. The next section will show you how to keep track of those observations.

# Keeping Track of Your Experiment

You can use the following charts to keep track of what occurs during your experiment, or you can make your own similar charts.

Use one chart to keep track of what happens during the first week, and another for the second week.

| | Day 1 | Day 2 | Day 3 | Day 4 | Day 5 | Day 6 | Day 7 | Location and Environment |
|---|---|---|---|---|---|---|---|---|
| Set 1 Bread | | | | | | | | Set 1 |
| Orange slice | | | | | | | | |
| Cheese | | | | | | | | |
| Set 2 Bread | | | | | | | | Set 2 |
| Orange slice | | | | | | | | |
| Cheese | | | | | | | | |
| Set 3 Bread | | | | | | | | Set 3 |
| Orange slice | | | | | | | | |
| Cheese | | | | | | | | |
| Set 4 Bread | | | | | | | | Set 4 |
| Orange slice | | | | | | | | |
| Cheese | | | | | | | | |
| Set 5 Bread | | | | | | | | Set 5 |
| Orange slice | | | | | | | | |
| Cheese | | | | | | | | |

*Keeping track of mold growth for the first week.*

Be sure to write down as many observations as you can each day. If you see a bit of mold growing on one piece of bread, for instance, and a lot of mold growing on another, be sure to note the location of the breads, the amount of mold and what it looks like, and anything else you might see.

Don't wait until the next day, or even later the same day, to write down what you see. Take the chart with you from room to room and write down what you see as you look at each bag.

|  | Day 8 | Day 9 | Day 10 | Day 11 | Day 12 | Day 13 | Day 12 | Location and Environment |
|---|---|---|---|---|---|---|---|---|
| Set 1 Bread |  |  |  |  |  |  |  | Set 1 |
| Orange slice |  |  |  |  |  |  |  |  |
| Cheese |  |  |  |  |  |  |  |  |
| Set 2 Bread |  |  |  |  |  |  |  | Set 2 |
| Orange slice |  |  |  |  |  |  |  |  |
| Cheese |  |  |  |  |  |  |  |  |
| Set 3 Bread |  |  |  |  |  |  |  | Set 3 |
| Orange slice |  |  |  |  |  |  |  |  |
| Cheese |  |  |  |  |  |  |  |  |
| Set 4 Bread |  |  |  |  |  |  |  | Set 4 |
| Orange slice |  |  |  |  |  |  |  |  |
| Cheese |  |  |  |  |  |  |  |  |
| Set 5 Bread |  |  |  |  |  |  |  | Set 5 |
| Orange slice |  |  |  |  |  |  |  |  |
| Cheese |  |  |  |  |  |  |  |  |

*Keeping track of mold growth for the second week.*

# Putting It All Together

Once you've got all your observations recorded on a chart or charts, you'll be able to look closely at what you've written and summarize what you've learned from your experiment.

When you've put together your results, you'll be able to see whether your hypothesis was correct. You will have reached a conclusion, which is the last step of the scientific method. When you reach a conclusion, you will have solved the problem you described at the beginning of your project.

Whether or not your results turned out to be what you thought they would is not the most important aspect of a science fair project. Having conducted the experiment and recorded your results properly is more essential than a correct hypothesis.

Once you've reached a conclusion to your experiment, the scientific part of your project is over.

# Further Investigation

If you want to take this project a step or two further, there are some easy ways to do so. You could leave some of the food enclosed—but not sealed—in bags to see if allowing air to enter the bags affects the growth of mold.

Put some of each type of food in sealed bags, and some of each type in unsealed bags in the same location and see what happens. Doing so will determine if mold grows better in a sealed or unsealed container, and, as in your original experiment, the best types of food and environments for growing mold.

**Standard Procedure**

For some really cool mold and fungi facts, check out this website: www.herb.lsa.umich.edu/kidpage/factindx.htm. It comes from the University of Michigan, and has some neat interactive activities and lots of fun information and photos.

To go another step further, use a vacuum packaging machine to remove all the air from the bag in which you place the different foods. These are available in kitchen stores and some department stores, or maybe there's already one in your home. They allow you to seal food in plastic without any air, therefore keeping out any microbes—such as mold spores—that can contaminate the food.

If you want, you can try adding some of these sealed bags to your sets and see what happens. Or you can think of your own ideas to further investigate how mold grows.

# Other Great Biology Projects

If you enjoyed experimenting with and learning about how molds grow, maybe you'd be interested in some other biology projects. Two other projects—one dealing with ants and the other with the human heart—are described in this section.

The descriptions aren't as detailed as the mold project, and we don't walk you step-by- step through the experiment. The information given, however, should be enough to get you started. Use the scientific method to prepare for and work through your experiment and to record the results.

## What Foods Attract Ants?

Did you ever wonder why ants seem to have a one-track mind for sweets? Leave a cookie on the counter in a kitchen that has become residence to some ants, and it's guaranteed that they'll find it—quickly.

So what is it with ants and their collective sweet tooth? Do ants even have teeth? No, they don't have teeth. They do, however, have a chemical sensory aid that directs them to the foods they love the most. It would be like you having a built-in radar device that beeps whenever you get within 50 yards of an ice cream shop.

In this science fair project, you try to determine which foods from your kitchen ants like best. The experiment is not conducted inside the kitchen (much to Mom's relief), but outside in a corner of your yard or other area in which you can work.

> ### Scientific Surprise
>
> Ants are pretty amazing little creatures. They can lift 20 times their own body weight. If a man could run at a comparable speed with an ant, he could run as fast as a racehorse. Alas, however, the average life expectancy of an ant is only 45 to 60 days.

To learn about the eating habits of ants, you can place several types of food in separate, flat container lids. Using a permanent marker, write on each lid the type of food it will hold. Be sure to include these three different types of foods:

- ◆ Sweets, such as candy, brown sugar, or fruit juice

- ◆ Proteins, such as meat, cheese, or milk

- ◆ Carbohydrates, such as crackers, or cooked rice or pasta

Every day for a week, place a small amount of various types of food in a shallow lid. You can use just one food from each group or more than one—it's up to you. Just be sure to include at least one of each type of food. And remember to put about the same amount of food in each lid. Group all the lids together in a small area accessible to ants.

Also, set out a lid containing plain water. That will serve as your control. The foods are the variables.

Once you've placed the food in the lids outside, you'll need to watch it pretty carefully. You'll want to take a look every couple of hours to see what's going on.

Take a notebook with you and record what you see—don't count on remembering later on. Try to establish patterns by watching which foods attract the most ants, and which foods the least. How long does it take for ants to find the food? How long do the same ants hang out at the outdoor café you've provided for them? Do more ants appear once the food site has been established for several days?

*Ants are attracted to different types of foods.*

When you've observed the behavior and habits of the ants over an entire week, you can chart your observations.

## Do Loud Noises Affect Heart Rate?

The human body is a marvelous creation. One of its really awesome features is its reflex system. *Reflexes* are sort of an automatic safety response system in your body, and they work to protect you from harmful situations.

Think about a time when you accidentally touched something hot, like the kitchen stove or an iron. Before you had time to think about what was happening, you had jerked your hand away from the hot surface. That immediate response was caused by a reflex.

> **Basic Elements**
>
> **Reflexes** are actions your body performs before your brain even knows what happened. They work to protect you from harmful situations, such as when you accidentally touch something hot.

The heat and pain sensors in your hand sent a message to the spinal cord, which sent a message to the arm and hand muscles. That message was something like, "Yikes, get me out of here!"

While all that was going on, a message was on its way to your brain, allowing pain to register after you had already pulled your hand away.

The brain is the control center of the human body. It communicates via chemical messages with the different systems that make up the body, such as the muscular, digestive, and circulatory systems. When your reflexes kick in, they activate the nervous system, which relays a message to the muscular system, telling it to "Move!"

Other things happen in your body as well when your reflexes are activated. Think back on some of your own experiences and think how you felt after you had a real scare or did something that put yourself in harm's way.

Can you remember feeling your heart pounding? Your pulse racing? Maybe your face felt very hot, or you even started shaking involuntarily. Your body reacts in many ways that are not within your control.

In this project, you're trying to find out whether loud noises affect the rate at which a person's heart beats. Form a hypothesis by thinking about some of your own experiences, and using your observations and research.

You'll need to get several people to agree to participate in your experiment. The first thing you'll do is determine the resting heart rate of each person with whom you're working. To do that, you'll need to locate the *pulse point* on each person's wrist. That's where you can feel the beating caused by regular contractions of the heart.

**Basic Elements**

The **pulse point** on a person's wrist is the spot at which you can feel the regular beating in the artery, caused by the contractions of the heart.

Record the number of beats you feel for 10 seconds, and then multiply that number by six. That will tell you how many times the person's heart beats in one minute, while at rest. Heart rate varies throughout the day in most people. A resting heart rate may be 70 beats per minute, while, during strenuous exercise, it could increase to 160 or 170 beats.

There are several ways you could conduct this experiment. One would be to get someone to be your assistant, and have him or her sneak up behind the person whose pulse you've just taken and drop a large, heavy book on the floor. The unexpected sound should be enough to startle your subject, after which you can immediately check the heart rate again to see if it's changed.

Check the person's pulse every minute for three minutes to see if the heart rate changes. If he or she was startled by the noise and the heart rate increased, perhaps it will decrease again as the effect of the noise wears off.

If you don't have someone to help you, you can make a loud noise yourself once you've established the resting heart rate. Be a little creative with this experiment.

This experiment will work best if you repeat it several times with different people as your volunteers. Check to see if you notice any patterns in the heart rates of the different people. Did everyone's heart react the same way? Was the age of the person a factor?

The more data you can collect concerning the effect of loud noise on heart rate, the more reliable your results will be. Repeating the experiment several times will give you the data you need to either prove or disprove your hypothesis.

## The Least You Need to Know

- ◆ Some molds and fungi are harmful, while others are beneficial.
- ◆ Learning about molds can help you to discourage their growth in and around your home.
- ◆ Considering what you already know can help you to make an educated guess concerning the results of your experiment.
- ◆ It's very important to note all your observations as soon as you make them.
- ◆ There are many ways in which you can expand on a science fair project or come up with a related project.
- ◆ Insects such as ants, and the human body, are other interesting areas of biology from which you can obtain science fair project topics.

<br />

# Chapter 8

# In Which Liquids Do Seeds Grow Best (and Other Great Botany Projects)?

## In This Chapter

- ◆ Identifying the problem you want to solve
- ◆ Formulating a hypothesis
- ◆ Gathering the materials you'll need
- ◆ Conducting your experiment
- ◆ Making sense out of your results
- ◆ Considering other botany projects

Normally, when we think about planting seeds, we think of planting them in the earth, or in soil that's been put into containers. A healthy seed placed into soil and given the proper amounts of water, light, and heat will sprout, or germinate, and grow into a plant.

But did you know that seeds also can be germinated in liquid? It's true. In fact, many plants can be grown into maturity without any soil at all. Growing plants in water instead of soil is called *hydroponics,* and it's a fascinating type of horticulture. Horticulture, as you might know, is simply the science of cultivating plants. Some people refer to horticulture as an art, particularly when it pertains to decorative plants.

For this science project, however, we're interested in seeing which liquids are best for sprouting seeds. You won't be using hydroponics to grow plants to maturity, but that's something you might want to explore on your own.

Let's get started now and find out which liquids seeds grow best in.

# So What Seems to Be the Problem?

The problem, or question that you'll answer in the course of this project, is "In Which Liquid Do Seeds Grow Best?" Following clearly defined steps and procedures, you'll solve the problem during the course of your experiment.

By the time you've finished your experiment, you'll know what type of liquid the seeds you've planted like the best—and the worst. Some seeds may grow quickly, and others may not grow at all. You will have answered whether seeds prefer milk, iced tea, vinegar, orange juice, club soda, or plain old tap water.

**Basic Elements**

Hydroponics is a fancy-sounding word, but it's nothing more than the practice of growing plants in a water solution to which nutrients have been added. The difference from the regular method of growing plants, of course, is that no soil is used.

The problem you've stated happens to also be the suggested title of your science fair project. While your problem must always be a question that you're attempting to answer, your science fair title does not necessarily have to be in question form. You also could use one of the following names as the title for this project:

 ♦ What's the Best Liquid for Germinating Seeds?

 ♦ Testing Variables When Sprouting Seeds

Once you've identified and stated your problem, you should take a few minutes to think about the point, or purpose of your project.

# What's the Point?

So why should you, or anybody else for that matter, give the slightest thought to sprouting seeds in different liquids? What's the point?

Until now, you've probably assumed that seeds sprout better in water than in any other liquids. Right? Well, maybe they do. In this project, water is the control, or the substance with which we'll be comparing all the other liquids. The other liquids are the variables, which you learned about in Chapter 5.

But, unless you've already experimented to find out, you can't be sure that water is the best liquid for sprouting seeds, can you? How do you know that a seed placed in orange juice won't grow into a beanstalk to rival the one Jack climbed in that famous fairy tale?

The point of this project is to use the scientific method to solve the problem stated above and find out in which liquids seeds grow the best.

**Standard Procedure** _____

If you're interested in learning more about hydroponics, check out a website called The Growing Edge. It's at www. growingedge.com/kids.

# What Do You Think Will Happen?

Before you state your hypothesis (see Chapter 4 for more on the hypothesis), take a few minutes to think about the liquids in which you'll be trying to sprout bean seeds. We chose bean seeds because they're fairly large and easy to work with, and they sprout quickly, usually in about a week.

Again, the liquids you'll be using are:

- ◆ Milk
- ◆ Iced tea
- ◆ Vinegar
- ◆ Orange juice
- ◆ Club soda
- ◆ Water

You've probably seen, smelled, and tasted nearly all of these substances. If not, you might want to do so before you come up with your hypothesis.

**CAUTION**

**Explosion Ahead** _____

*Never, never, never* put anything in your mouth or on your skin unless you're absolutely, one hundred percent sure that it won't hurt you. None of the materials used in this project will cause you any harm, but you may want to use caution if you're going to smell or taste vinegar— it can burn if it gets in your eyes.

Once you've examined the liquids you'll be using, try to think of some ways in which they might affect the seeds. Are some of the liquids more nourishing than others? Do you think any of them might actually harm the seeds? Once you've given the matter some thought, you can make and record your hypothesis.

# Materials You'll Need for This Project

Everything you need for this project should be easy to find. In fact, you probably already have most of the materials in your house. If your science fair is in the middle of winter and you live in a cold climate, you might have a little trouble finding bean seeds in a store near your home. Everything else, however, should be readily available. You need:

♦ 8 ounces (240 ml) each of tap water, milk, iced tea, vinegar (either white or cider), orange juice, and club soda

♦ One package of bean seeds

♦ Six (10- or 12-ounce) (300 to 350 ml) glass or plastic cups, all the same color and size

♦ Permanent marker

♦ Tray or shallow pan

♦ Metric ruler

♦ Paper towels

**Standard Procedure**

If you can't find bean seeds in a store near your home, there are lots of Internet sites from which you can purchase them. One of those sites is the Garden Shop Online, found at gardenshoponline.com. A pack of bean seeds will cost you about $2.50.

Be sure that you've cleared a space and have everything you'll need for your experiment before you get ready to start. And always be sure to check with a parent before you use household items such as glasses or markers.

# Conducting Your Experiment

Once you've completed the early steps of the scientific method as explained in Chapter 4, you'll be ready to start your experiment.

Remember to work carefully and in a logical manner. Try not to knock over any of the glasses, and be sure that you write the correct names of the liquids on the glasses. Water, club soda, and white vinegar all look pretty much the same, but they may have very different effects on bean seeds.

1. Using the permanent marker, label each of the six cups with the name of the liquid it will contain.

2. Place the cups on the tray or shallow pan.

3. Pour 8 fluid ounces (240 ml) of water (your control liquid) into the cup labeled *water.*

4. Pour 8 fluid ounces (240 ml) of each of the other liquids into its proper cup.

5. Open the package of seeds and divide them evenly into six piles. You might have a few seeds left over that you won't use.

6. Slowly add the seeds from the first pile into the cup labeled *water*.

7. Continue putting the other piles of seeds into each of the five remaining cups.

8. Place the tray with the cups where you can easily observe the seeds. You'll want to keep the temperature as constant as possible, so make sure the seeds are in an area where there are no drafts. And, make sure the seeds won't get bumped or knocked over.

9. Keep a daily record of your observations. To do so, write down how many seeds are in each cup, and how many seeds break their shells and begin growing, or germinating, every day. Use the first chart shown in the next section, "Keeping Track of Your Experiment," to help you to record your observations.

**Standard Procedure**

Remember that there are different indicators that will help you figure out which liquid the seeds like best. At this point of the experiment you're looking to see how many seeds in one glass sprout, and which seeds sprout the most quickly. Those are two indicators that will help you determine your results.

*The stages of seed germination.*

**Standard Procedure**

Be sure to measure all the seedlings from one cup and replace them into their proper cup before starting to measure those that were sprouted in a different liquid. It's absolutely necessary that you keep the sprouts in their proper cups.

10. A week after you placed the seeds in the various liquids, measure each sprout in centimeters. If your seeds haven't sprouted yet, sit tight and begin measuring them at two weeks. You'll need to record the length (in centimeters) of each sprout in every cup. You can make it easier to measure the sprouts by removing each one, placing it on a paper towel, and measuring its length with a metric ruler. Once you've measured every sprout, you'll need to figure out the average length of the sprouts in each liquid.

Use the second chart found in the next section to record the average length of sprouts in each cup after one or two weeks of growth. The average length of the sprouts in each cup is another indicator of which liquid the bean seeds like the most.

Follow these steps to calculate the average length of all the sprouts in each of the liquids:

Add the lengths of all the sprouts from one cup. If you had four beans in one cup and each of the sprouts was ½-inch long, for instance, your total would be 2 inches.

Divide the total length by the number of sprouts you measured.

The average length equals the total length divided by the number of sprouts measured.

# Keeping Track of Your Experiment

Once you've observed and recorded everything that occurred during the course of your experiment, it's time to present this information clearly.

To do this, you'll need to present all of the measurements you've taken. These measurements should include the following:

♦ The number of seeds you placed in each liquid

♦ The number of seeds in each liquid that began germinating on a daily basis

♦ The average length of seedlings in each liquid after a one- or two-week period

All of these measurements should be neatly presented on the following charts. The number of seeds placed in each liquid and the number of seeds that germinated in each liquid should be recorded on the first chart. The average length of the seedlings should be recorded on the second chart.

**Daily Observations of Seeds**

| | Milk | Iced Tea | Vinegar | Orange Juice | Club Soda | Water |
|---|---|---|---|---|---|---|
| Number of seeds in each cup | | | | | | |
| Day 1 # of seeds germinating | | | | | | |
| Day 2 # of seeds germinating | | | | | | |
| Day 3 # of seeds germinating | | | | | | |
| Day 4 # of seeds germinating | | | | | | |
| Day 5 # of seeds germinating | | | | | | |
| Day 6 # of seeds germinating | | | | | | |
| Day 7 # of seeds germinating | | | | | | |
| Day 8 # of seeds germinating | | | | | | |
| Day 9 # of seeds germinating | | | | | | |
| Day 10 #of seeds germinating | | | | | | |

*Use this chart to record your daily observations of the seeds.*

**Length of Sprouts in Centimeters**

| Sprout Number | Milk | Iced Tea | Vinegar | Orange Juice | Club Soda | Water |
|---|---|---|---|---|---|---|
| 1 | | | | | | |
| 2 | | | | | | |
| 3 | | | | | | |
| 4 | | | | | | |
| 5 | | | | | | |
| 6 | | | | | | |
| 7 | | | | | | |
| 8 | | | | | | |
| 9 | | | | | | |
| 10 | | | | | | |

*Use this chart to record the length of each sprout, in centimeters.*

# Putting It All Together

Once you've analyzed your data, you'll be able to summarize what you've learned, and you'll see whether your hypothesis was correct. You will have reached a conclusion, which is the last step of the scientific method. You will have answered the question posed in your problem, and know if you were right in making your hypothesis.

Remember that if your hypothesis turned out to be incorrect, it doesn't mean that your experiment was a failure. It just means that the results you got were not the ones you thought you would.

Part 6 of this book tells you all about great ways to display your project and write any reports or other papers that may be necessary. The experiment part of your project, however, has ended.

# Further Investigation

If you enjoyed doing this project and would like to take it a step or two further, consider what would happen if you grew the seeds in regular potting soil, but watered them with the different liquids you used for this experiment.

Maybe something in the potting soil would react favorably with orange juice, for instance, and cause the beans to shoot up in record time. Perhaps if you watered the seeds with Hawaiian Punch, your bean plants would produce red beans.

If you're interested in investigating this idea, just adapt the steps you used when sprouting beans in the liquids. And have fun!

**Standard Procedure**

Nearly every scientific experiment can be adapted or modified in order to solve a similar, but not entirely the same, problem. Don't be afraid to use your scientific curiosity and think of other questions for which you want to discover answers.

# Other Great Botany Projects

If you enjoyed working with seeds and seedlings, you might be a botanist at heart. A botanist is someone who studies and is knowledgeable in the field of *botany*—or plant science.

A couple other projects involving the scientific area of botany are described next. The information presented will give you a general idea of how to do the project, but doesn't walk you through all the steps like with the earlier project described in this chapter. Don't forget that you'll need to follow the steps of the scientific method.

**Basic Elements**

**Botany** is the science dealing with the structure of plants, the functions of their parts, the conditions in which they grow, and the language used to describe and classify them. A botanist is a scientist who studies plants and related subjects.

# Does Microwave Radiation Affect Seed Growth?

The title of this experiment might sound a little scary, but it actually just means that you'll be testing what happens to seeds that are placed in a microwave oven and "zapped."

The point of the project is to learn whether seeds that have been exposed to microwave radiation (that's the stuff that occurs when the microwave is turned on) grow as well, or better, than seeds that haven't been "zapped."

> ### Scientific Surprise
>
> A microwave is a type of energy in the form of electromagnetic radiation. Electromagnetic radiation includes cosmic ray photons, gamma rays from radioactive elements, x-rays used in hospitals, UV light waves from the sun, visible light from a light bulb, infrared radiation, microwaves, radio waves, and electric currents.

All you need to do is put equal numbers of bean seeds (or other seeds, but beans are easy to work with) on damp paper towels, then heat each batch for a different amount of time in your microwave oven. The amount of time to which the seeds are exposed to microwave radiation should vary in 30-second intervals from 30 seconds to three minutes. For instance, you'd put the first batch of seeds in the microwave for 30 seconds, the second batch for a minute, the third batch for one minute and 30 seconds, and so forth.

After all the beans (except the control group) have been "zapped," you'll plant them in the same type of containers and soil. Make sure that all the containers are the same size, and the seeds all receive the same amounts of light and water.

> ### Scientific Surprise
>
> When you press the START button on your microwave oven, the oven generates microwaves from the electrical current flowing through the power cord that is plugged into your wall. These microwaves are able to be absorbed by water, fat, and sugars—the main components of the foods we eat. When the microwaves penetrate the food, they excite the molecules in the food and cause them to move very quickly. The movement of the molecules causes the food to quickly heat up all over.

After two weeks, during which you'll record your observations, measure every sprout and determine the average length of plants in each container, just as you did when sprouting beans in the various liquids. Your results will answer the question of whether or not microwave radiation affects the growth of bean plants.

## Which Bulbs Are Best for Forcing?

You've probably seen tulips, daffodils, crocuses, lilies, hyacinths, and other bulbs blooming outside in the spring and summer. The bulbs from which these flowers grow lie dormant all winter, then sprout in the spring or summer when it's time for them to grow.

By altering the biological clocks of such bulbs, however, you can force them to open in the winter, before they normally would. For this project you'll need several different types of flower bulbs, potting soil, trays, and same-size containers.

*Flower bulbs before shoots appear.*

Plant several of each type of bulb in a pot, being sure to leave uncovered the part of the bulb from which the leaves appear. The broader part of a bulb is generally the bottom, and is the side from which the roots grow. That part should be placed in the dirt so that the roots can grow downward. Once you've planted five or six different types of bulbs, move the pots to a dark, cool place, such as a garage or unheated basement. The temperature should ideally be at about 38 degrees Fahrenheit.

This project requires a good amount of patience and advance planning, for it can take many weeks for bulbs to begin to grow. Once they have, however, and are actively growing, move the pot to an area that receives indirect sunlight and is about 55 degrees Fahrenheit. As the white tips of the plants turn green, increase the amount of

**Standard Procedure** _____

Some bulbs that generally are fairly easy to force indoors are daffodils, narcissus, tulips, crocus, hyacinth, grape hyacinth, and iris.

sunlight and the temperature. When the buds appear to be ready to open, you can move them to an area that is warmer and sunny.

Rates of growth may vary tremendously between the different types of bulbs, and some may not respond to forcing at all. Just be sure that you use the exact same procedure for each type.

There are thousands of science fair projects that deal with botany, many of which you can probably think of on your own. Just identify a problem relating to plants, and use the scientific method to reach a conclusion.

## The Least You Need to Know

◆ You need to identify a problem or question before preparing for a scientific experiment.

◆ Assessing properties of the materials you'll use may help you to formulate your hypothesis.

◆ Most of the materials required for this project are common household items or items readily available from a variety of sources.

◆ It's extremely important to work carefully during your experiment and to stay well organized.

◆ Properly measuring and presenting your results are very important in putting together a great science fair project.

# Salt or Sugar: Which Dissolves Faster in Different Liquids (and Other Great Chemistry Projects)?

## In This Chapter

- Understanding what makes a solution
- Factors affecting the rate at which materials dissolve
- Knowing your hypothesis may not be correct
- The importance of patience when doing your experiment
- Figuring out your results
- Expanding the experiment for further learning

Solutions are nothing more than mixtures of different compounds or elements. You encounter solutions every day without even realizing it.

Even the air you breathe—which contains water—is a solution of a liquid and a gas. If you drank a soda today, you actually drank of solution of a gas dissolved in flavored water. If you're wearing a bracelet made of sterling silver, you're wearing a solution of two metals.

In this experiment, you'll be working with a liquid solution, which is one of three types of solutions. The other types are gaseous solutions and solid solutions.

# So What Seems to Be the Problem?

When you stir a spoonful of sugar into a glass of water, you are forming a solution. This type of liquid solution is composed of a solid solute, which is the sugar, and a liquid solvent, which is the water. As the sugar molecules spread evenly throughout the water, the sugar dissolves.

Mixing a liquid in a gas makes another type of solution, called a gaseous solution. An example of this type of solution is humidity. Humidity is water (a liquid) dissolved in air (a gas).

### Basic Elements

A **solute** is the substance—either a solid, liquid or gas—that gets dissolved. A **solvent**—which also can be a solid, liquid, or gas—is the substance that does the dissolving. A **solution** is a uniform mixture of a solute (usually a solid) dissolved in a solvent (usually a liquid).

### Standard Procedure

Think about how a sugar cube dissolves in water, compared to a package of loose sugar. The cube dissolves more slowly because fewer sugar molecules are initially in contact with the water.

In a solid *solution*, such as sterling silver, copper that has been heated at high temperatures is mixed with silver that also has been heated until it melts. The copper is the *solute*, which is the substance that will dissolve into the *solvent*. The silver is the solvent.

The type of solution is determined by the state of matter of the solvent. If the substance doing the dissolving is a liquid, the solution is called a liquid solution. If the solvent is a gas, the solution is called a gaseous solution. And you guessed right: A solid solvent will form a solid solution.

There are a few factors that generally increase the amount of solute that can be dissolved. If you want to dissolve more sugar in the same amount of water, for instance, you could heat the water. You also could grind the sugar into smaller particles to increase its surface area, or you could stir the mixture.

In the years that you've been using salt and sugar on your foods, you've probably noticed that each piece of salt—which actually is a crystal—is a little smaller than each piece of sugar, which also is a crystal.

**Standard Procedure**

You can check out the size of salt and sugar crystals under a microscope or magnifying glass, which will allow you to see their shapes, as well. If you draw what you see, using a pencil so that you can illustrate shading, you could include the illustration as part of your final display of your science fair project.

The problem you'll be attempting to solve in this experiment is whether sugar or salt dissolves faster when mixed into various liquids. Does the size of the pieces affect how quickly they mix with the liquid?

When you dissolve sugar or salt in a liquid— say, water—what happens is that the sugar *molecules* move to fit themselves between the molecules of water within a glass or beaker. The illustration below shows how the different molecules are arranged in the container.

**Basic Elements**

A **molecule** is two or more elements that are chemically combined. A molecule of salt contains sodium and chlorine, which are chemically combined to make sodium chloride. The chemical formula for this salt is NaCl. A molecule of sugar contains carbon, hydrogen, and oxygen. The chemical formula for sugar is $C_{12}H_{22}O_{11}$.

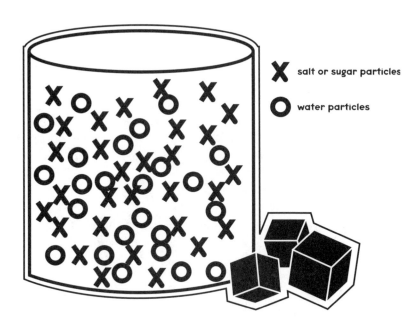

**X** salt or sugar particles

**O** water particles

*A solute, such as sugar, dissolved in a solvent, such as water, results in a liquid solution.*

In your experiment, you'll see how salt and sugar molecules move within different liquids and dissolve at different rates.

The title of this chapter, "Salt or Sugar: Which Dissolves Faster in Different Liquids?" could serve as your project title, if you want. You also could consider one of the following titles for your project:

◆ The Great Salt vs. Sugar Dissolving Contest

◆ Using Salt and Sugar to Explore How Substances Dissolve

Whatever name you choose is fine. Let's take a minute now to consider why this project is a valuable use of your time.

# What's the Point?

The point of this experiment, in addition to learning whether salt or sugar dissolves faster in various liquids, is to learn how molecules interact in a solution.

As you saw in the preceding illustration, the water molecules take up most of the room in the container. But there is still some available space in which the sugar or salt molecules can fit. Through your experiment, you'll learn how fast the sugar molecules fit into those spaces, as compared to the salt particles.

Knowing this will help you better understand the process that occurs as a substance dissolves.

---

**Basic Elements**

When there is ample space between the molecules of a solvent, the solvent is said to be **unsaturated**. When a lot of solute has been dissolved in the solvent, but there is still some space between the molecules, the solution is **concentrated**. When absolutely no more solute can be dissolved within a solvent, the solution is **saturated**.

And when the excessive solute has been dissolved by heating the solution, it's said to be **supersaturated**.

---

The control in your experiment (if you need a refresher on controls and variables, see Chapter 5) will be water. The other liquids in which you dissolve salt and sugar will be the variables.

|  |  |
|---|---|
| Controls: | Solvent—water |
|  | Solutes—sugar, salt |
| Variables: | Five different clear liquids (may be colored) |

Remember when you conduct your experiment that it's very important that the liquids you use are all the same temperature. You already learned that sugar dissolves faster in a warm liquid than in a cool one, so you know it wouldn't be an accurate experiment if some of the liquids you use are warm and some are cold. The temperature of the liquid would become a variable.

Therefore, all the liquids you use—including water—should be at room temperature. If you normally keep them in the fridge, be sure to allow them to sit out on the counter overnight until they are all the same temperature.

**Explosion Ahead**

Don't assume that liquids that have been sitting in different areas of your house are the same temperature. A bottle of soda that's been sitting in the garage, for instance, may be several degrees cooler than rubbing alcohol from the bathroom closet, or apple juice from the kitchen pantry. Be sure to have all liquids in the same location in order for them to achieve the same room temperature. If you don't, the results of your experiment won't be valid.

To give you a little more flexibility when you conduct the experiment, you may choose the liquids in which you'll dissolve sugar and salt. There's no point in having to go out and buy additional liquids if you've already got what you need.

Just make sure you choose liquids that are different from each other in taste, color, odor, and purpose. You'll also need to select those that allow you to observe the salt and sugar as it dissolves. If you use milk or orange juice, for example, you won't be able to watch the salt and sugar dissolve. Some suggestions for liquids to consider are:

- White vinegar
- Club soda
- Ginger ale
- Glass cleaner (such as Windex)
- Lemonade
- Tea or iced tea (each at room temperature)
- Apple juice
- Rubbing alcohol

All of these are commonly found around the house, perhaps saving you a trip to the store.

# What Do You Think Will Happen?

Now that you know how solutions are formed and some of the factors that will affect the speed at which the sugar and salt you'll be using will dissolve, you should be able to make a good guess as to which one will dissolve faster.

**Standard Procedure**

If you haven't done this experiment before, you won't know if the liquids you use will be a factor in dissolving salt and sugar. This makes it more difficult to form a hypothesis, but don't worry. Whether your hypothesis turns out to be correct, or not, does not affect the validity or outcome of your experiment.

While you won't know until after your experiment if properties of the different liquids you choose will affect the rate at which the salt and sugar dissolve, you do know that salt crystals are generally smaller than sugar crystals. And you know that the temperature of the liquids will not be a factor in your experiment.

Just try to use your past experiences, the information you've read earlier in this chapter, and your common sense to come up with a sound hypothesis.

Remember that your hypothesis must be stated as an objective sentence, not a question. So go ahead and make your guess as to whether the salt or sugar will dissolve faster, and let's get started with the experiment.

# Materials You'll Need for This Project

Some liquids suggested for use in this experiment are white vinegar, club soda, ginger ale, glass cleaner, rubbing alcohol, apple juice, lemonade, and tea. If you want to substitute another liquid for one or more of the ones suggested, that's fine. Just be sure that all liquids are clear and at room temperature.

The amounts of materials listed below are enough for you to conduct the experiment three times with each liquid. You'll need:

◆ 12 clear, plastic cups (10 ounce [300 ml])

◆ One permanent marker

◆ One (1 teaspoon) (5.0 ml) measuring spoon

◆ One (½ teaspoon) (2.5 ml) measuring spoon

◆ One (1 cup) (240 ml) measuring cup

◆ 8 teaspoons (40 ml) salt, divided in 16 (½ teaspoon) portions

- 8 teaspoons (40 ml) sugar, divided in 16 (½ teaspoon) portions

- 48 ounces (1,440 ml) water at room temperature

- 24 ounces (720 ml) each of five different, clear liquids, all at room temperature

- One clock or watch with a second hand

- One clear plastic cup containing eight fluid ounces (240 ml) water at room temperature

Remember to make sure that all liquids are at room temperature.

# Conducting Your Experiment

When you've gathered all your materials, you'll be ready to begin your experiment. Just follow these steps:

1. Using the permanent marker, write "salt" on six of the plastic cups, and "sugar" on the other six.

2. Place ½ teaspoon (2.5 ml) of salt into each of the six cups labeled "salt."

3. Place ½ teaspoon (2.5 ml) of sugar into each of the six cups labeled "sugar."

4. Add 8 ounces (240 ml) water to one cup containing salt, and one cup containing sugar. Immediately record the time at which the water was added on a data chart similar to the one shown in the next section, "Keeping Track of Your Experiment."

 **Standard Procedure** _____

Keep a cup of plain water in sight so you can compare it to the cups containing salt and sugar. It will be interesting to watch how the appearances of the liquids change as the salt and sugar dissolve.

5. Observe the solutes (salt and sugar) dissolving in the solvent (water). Record on the data chart the time at which it appears to you that each solute has completely dissolved. These times will probably not be the same.

6. Calculate the elapsed time during which the dissolving occurred. Take the time at which the water was added to the cups and the dissolving started, and subtract it from the time the dissolving ended. This gives you the total minutes it took for the salt and sugar to completely dissolve in the liquid.

CAUTION

**Explosion Ahead** _____

It's going to take a little while for the sugar and salt to dissolve. For best results, *do not* stir the solutions, as doing so will present an additional variable. If you *must* stir, then stir each solution three times, and stop. Only stir after you notice there is solute at the bottom of each of the two containers. Stirring the solutions unevenly will cause your experiment to be invalid.

7. Repeat steps 4 through 6, using each different liquid instead of the water.

8. Wash, rinse, and thoroughly dry each of the 12 cups.

9. Repeat steps 2 through 8 two more times, for a total of three trials for each of the six liquids.

10. Calculate an average dissolving time for the salt and the sugar in each of the six liquids.

Remember that to find the average time it took for the salt and sugar to dissolve in each liquid, you add the three times recorded for each one, then divide them by three. The number you get when you divide the times is the average time.

# Keeping Track of Your Experiment

Charts such as the following one can be used to record information for each solvent. Simply change the names of the solvents in the heading.

**Water**

|  | Time Started | | | Time Ended | | | Time Elapsed | | |
|---|---|---|---|---|---|---|---|---|---|
|  | Trial | | | Trial | | | Trial | | |
|  | 1 | 2 | 3 | 1 | 2 | 3 | 1 | 2 | 3 |
| Salt |  |  |  |  |  |  |  |  |  |
| Sugar |  |  |  |  |  |  |  |  |  |

*Use this chart to record the time it takes for the sugar and salt to dissolve.*

Be sure to record the times as you go along. Don't depend on your memory to write them down later. You're going to have a lot of numbers by the time you finish your experiment.

# Putting It All Together

What did you notice about the rates at which the salt and the sugar dissolved? Did you prove your hypothesis to be correct? Or incorrect? Could you detect any type of pattern as you added the salt and sugar to the various liquids? Was it obvious that the salt dissolved better and faster in some liquids compared to the sugar? Can you think of any reasons for which that might have occurred?

Do you think that the chemical natures of the solute and the solvent affected the dissolving rates? Use the information you gathered when you researched your topic to help you answer these questions.

The more you know about your project, the better able you'll be to analyze your data correctly and come up with a sound conclusion.

# Further Investigation

As mentioned earlier, the factors affecting the solubility of solid solutes are:

- Increasing or decreasing the temperature of the solvent
- Increasing the surface area of the solute
- Stirring

If you wanted to take this project a step or two further, you could design an experiment that would test one—or perhaps all—of these variables.

You could easily compare the rate at which sugar cubes dissolve in liquid with the dissolving rate of granulated sugar.

Or you could use the same solute—say, sugar—and test whether stirring the solution caused it to dissolve faster. Heating and cooling the solvent as you add the same solute also would be a possibility for further experimentation.

If you're curious and willing to experiment, you probably can think of many variations for this project. And, because the experiment calls for only common, inexpensive materials, you should be able to experiment to your heart's content.

# Other Great Chemistry Projects

If you enjoyed doing the experiment outlined in this chapter, perhaps you'll be interested in learning a little more about chemistry by working with the projects described in the following text.

One project tests the results of various household cleaning products on Jell-O. The other project tests some substances to see which is most effective at melting ice.

## What Household Cleaners Break Down Jell-O Best?

Before you start wondering why anyone would waste his or her time comparing the effects of household cleaners on Jell-O, think about the fact that Jell-O contains gelatin, a source of protein. So what, you say?

Gelatin is a protein formed by boiling specially prepared animal skin, bones, and connective tissue. It's used in some foods, drugs, and photographic film.

Our bodies contain enzymes that help to break down the protein we eat. When you eat meat, fish, cheese, eggs, yogurt, or other foods that contain protein, these enzymes get busy, working to make the proteins small enough to be useful to your body. The enzymes reduce proteins by breaking their chemical bonds. There are different kinds of enzymes, but the ones that specifically target proteins are called *proteases*.

While our bodies contain enzymes, so do some household cleaners. Enzymes in cleaning products work to break down protein-containing stains, such as food, milk, or blood.

By using five different household cleaners (pick whichever ones you please, but try to include cleaners used in different parts of the house, such as bathroom, kitchen, laundry, and so forth) as variables and distilled water as your control, you'll be able to see which products most effectively break down protein contained in the Jell-O.

Begin your experiment by preparing a batch of Jell-O, according to package instructions. Instead of putting it into a bowl to set, however, pour the hot

**Basic Elements**

**Proteases** are a specific type of enzyme that targets proteins. They work to break down proteins so that they're small enough for the body to use.

**Standard Procedure**

If you have a camera available, be sure to take some "before" and "after" pictures of your experiment. Take a picture just after you've added the cleaners to the Jell-O, and another one the following day. These photos will give you a great way to illustrate your experiment.

liquid into the sections of an ice cube tray. Be sure to get permission before you do this and ask for help if you need it.

Once the Jell-O has cooled and solidified, use a sharp knife to cut out a little hole in the middle of each section. You'll have to remove a small plug of Jell-O to make the hole.

Fill the holes in two sections of Jell-O with one cleaner. Fill two more holes with a different cleaner, and so on. Be sure to label which cleaners you put into each section, and be sure to add distilled water—your control—to two sections.

Allow the Jell-O and cleaners to sit undisturbed overnight, and observe them carefully the next day. This will require that you keep your experiment out of reach of your little brother—and your dog. Record your observations of all the sections, and don't forget to take a picture.

If one cleaner broke down the Jell-O much more than another, compare the ingredients of the cleaners and see if you can figure out what might have caused the difference. Try to do some reading about proteins and enzymes so you'll be able to complete your project and analyze your results.

**Standard Procedure**

To learn more about proteins and enzymes, check out the Cool Science for Curious Kids website at www.hhmi.org/coolscience, or the Science Made Simple website at www.sciencemadesimple.com.

## What Substance Melts Ice Fastest?

Another fun chemistry project is to test a variety of products to see which one works best to melt ice. As you probably know, materials are routinely used on roadways after snow and ice storms to provide traction and reduce slippery conditions.

Unfortunately, most of these applied chemicals can cause damage to soil and pollute water in lakes. It's important that we keep working to come up with alternatives to these chemical products that won't harm the environment.

In this experiment, you'll sprinkle various materials on ice cubes to test their melting properties. The variables you'll use are the different types of products, and the control is nothing. That is, you'll leave some ice untreated to see if it takes longer to melt than that which has been treated.

Substances you could use include road salt, calcium chloride, fertilizer, sand, and unused cat litter. All of these can be purchased at a home improvement or discount store.

> **Scientific Surprise**
>
> Cities and towns use hundreds of thousands of tons of salt and sand a year to keep the water on the roads from freezing. The salt, known as calcium chloride, helps to melt ice and snow by lowering the temperatures at which freezing can occur. The problem is that some of that salt and sand could end up in drinking water and cause health problems. Finding an economical alternative to road salt would be an earth-friendly solution.

To begin, empty a tray of ice cubes into a bowl. Sprinkle 12 ounces (336 grams) of one of the ice-melting materials over the ice cubes, then observe how long it takes for all the ice cubes to melt.

Write the elapsed time on a data chart. Repeat the procedure using a new set of ice cubes, a different product, and a clean, dry bowl.

When you've tested each product and also melted a tray of ice cubes by simply letting them sit at room temperature, you can compare differences in the melting times.

Did any of the materials come close to melting the ice as quickly as the road salt? What other materials might you use in this experiment? Can you think of any substance that might be a good replacement for the commonly used road salt?

## The Least You Need to Know

- Solutions are simply mixtures of different compounds or elements.
- There are solid solutions, gaseous solutions, and liquid solutions.
- Smaller particles generally dissolve faster than larger ones, and most materials dissolve quicker in a warm solvent than a cool one.
- A variety of liquids can be used in this experiment, as long as they are clear and the same temperature.
- The experiment may take some time to complete, and it's important that you be patient.
- Many variations can be added in order to vary the experiment.
- There are many fun, easy experiments that relate to chemistry.

# What Kind of Trash Bag Breaks Down Fastest (and Other Great Earth Science Projects)?

## In This Chapter

♦ Understanding how different materials break down

♦ Using your experience to generate a hypothesis

♦ Comparing paper and different types of plastics

♦ Building your "landfill" and conducting your experiment

♦ Making sure your observations are complete

♦ Other earth science experiments to consider

Americans produce a lot of trash, there's no question about it. The amount of stuff we throw away in this country is staggering.

The U.S. Environmental Protection Agency estimates that Americans generate about 4.6 pounds of trash per person—every day. Forty years ago, each person produced only 2.7 pounds each day.

There are nearly 300 million people in the United States. You do the math. Is it any wonder our landfills are filling up faster than we can figure out what to do about it?

As you can imagine and probably know, trash is a weighty (no pun intended) topic in this country. With only so much landfill space available, scientists and environmentalists are looking to other means of disposing of trash, such as burning it. We all, however, can help cut back on the amount of trash we generate by keeping in mind the motto of environmentalists—reduce, reuse, and recycle.

In addition to the amount of trash we produce as individuals and as a country, how we dispose of it is another problem. The plastic bags that most of us place outside of our homes to be hauled off to landfills every week are not exactly what you'd call environmentally friendly.

Nobody knows for sure how long regular plastic bags take to totally degrade, or break down, in a landfill, because we've only been using plastic commercially for about 90 years. It's estimated, however, that it may take 100 years for a plastic bag to completely degrade. It's sort of depressing to think that a plastic bag can outlive the majority of people on the planet.

In this chapter, we'll look at several types of bags and try to determine which ones break down faster when dumped into a landfill. Once you know, you can become an environmental ranger and start encouraging everyone around you to use the most environmentally friendly bag.

# So What Seems to Be the Problem?

The problem on a large scale, as stated above, is that we generate too much trash in this country, and much of it is material that will take many, many years to degrade.

**Standard Procedure**

If you're thinking of doing this project, remember that you need to begin the experiment eight weeks in advance of your science fair due date.

To solve this problem, we've got to learn to produce less trash, and to produce trash that won't stick around so long.

The problem you'll be attempting to solve in the course of this science fair project is related to the second part of the equation above. In addition to producing less trash, we've got to cut the time that it takes for our trash to degrade. Your task in this project is to test different types of trash bags, and determine which type of bag is the most *biodegradable*.

*This symbol means that a product or package may be recycled.*

It makes sense, doesn't it, to use trash bags that break down as quickly as possible, thereby allowing whatever is inside of them to degrade, as well.

Some manufacturers have come up with plastic bags that they claim are biodegradable, and some people use paper bags to hold their trash. Most folks, however, happily fill up one plastic trash bag after another, causing our landfills to be inundated with the long-lasting material.

In this project, you'll test how fast the following types of bags degrade when they're buried in the ground:

- ◆ Paper grocery bag
- ◆ Plastic grocery bag
- ◆ Standard trash bag
- ◆ Plastic trash bag labeled as *biodegradable*

This experiment will give you an idea of how fast these bags might break down in a landfill.

If you want, you can use the title of this chapter, "What Kind of Trash Bag Breaks Down Fastest?" for the title of your science fair project. Other titles you might consider are:

- ◆ Paper or Plastic, What's Best For the Earth?
- ◆ Best Bets for Bagging Your Trash
- ◆ Making a Biodegradable Choice

**Basic Elements**

The U.S. Environmental Protection Agency's definition of **biodegradable** is simply "capable of decomposing under natural conditions." Nothing difficult about that.

**Scientific Surprise**

Of all the trash (also called municipal solid waste) generated in this country, more than 37 percent of it is paper. Yard trimmings such as branches, leaves, and grass clippings make up 12 percent of the total, food scraps 11 percent, plastics nearly 11 percent, and metals about 8 percent.

Once you've chosen a title for your project, let's move along and have a look at what the purpose of such an undertaking might be.

# What's the Point?

The environment is a hot topic these days, and kids are getting involved in all sorts of projects and movements to help save it.

**Explosion Ahead**

Don't even think about starting this project until you get the okay from your parent or guardian. Trust me on this, very few adults will be amused to come home from work and find that you've dug a "landfill" in the backyard. Besides, you may require some help with the digging.

There are groups of kids and young adults working to save rainforests, animals, and waterways. Kids have formed recycling awareness groups in their schools and neighborhoods, often serving as a community conscience.

Seeing firsthand what happens to trash bags when they're buried will give you knowledge and help you to make informed decisions about what type of bags are best for you and your family to use for your trash. Once you know how the different bags degrade, you can tell others. You could even start a movement in your neighborhood to use bags that are easier on Mother Earth.

When you conduct this experiment, the supermarket brown paper bag will serve as your control. The other bags—the plastic grocery bag, the standard heavy-duty, black trash bag (such as Hefty brand), and a plastic trash bag that's marketed as being biodegradable (such as EcoSafe)—are your variables.

> **Scientific Surprise**
>
> Never think that kids can't make a difference. A nine-year-old girl named Melissa Poe started a group called Kids F.A.C.E. (For A Clean Environment) in 1989 in Nashville, Tennessee. The club had six members. As word spread about the organization, Melissa appeared on TV shows and there were stories about her in newspapers and magazines. Today, Kids F.A.C.E. has 300,000 members in 15 different countries. Never think that kids can't make a difference.

# What Do You Think Will Happen?

You may already have a strong feeling about what will happen to different kinds of bags when they're buried in your yard. You've probably handled enough paper and plastic to know what happens when they get wet, or how they withstand heat and cold.

From reading or hearing about environmental problems associated with plastic, you probably have the idea that paper is more environmentally friendly. But, what about plastic that is made to be environmentally friendly? Do you have an idea how that might compare to paper? Or to regular plastic? Do certain types of plastics break down more readily than others?

Think about one of those large, black plastic trash bags that you see sitting out on the curb on trash pickup day. Have you ever loaded one up and carried it outside? They're designed to hold a lot of trash, and the plastic they're made of is fairly thick and heavy.

How do you suppose that type of plastic bag compares in the environmentally friendly department with one of the plastic bags in which you carry groceries home from the store?

**Standard Procedure**

An easy way to cut down on trash is to buy reusable cloth bags in which to carry groceries. It makes no sense to buy hamburger rolls in a plastic bag, carry them home in a plastic shopping bag, and then throw both bags away. Remember the three Rs: reduce, reuse, and recycle.

Take a little time to think about these questions, and about the experiences you've had with paper and plastics. Then go ahead and make a smart guess, or a hypothesis, about which type of bag will break down the fastest, and which will take the longest.

# Materials You'll Need for This Project

All you'll need to conduct this science fair experiment are the four types of bags mentioned previously, a few other objects, and some solid municipal waste—also known as trash or garbage.

You really can use whatever garbage you want inside your trash bags, but it must be the same objects and amounts in each one.

Here's a list of materials you'll need, and a suggested list of solid municipal waste. Feel free to adapt it to whatever you can find to use around your house.

- One brown, paper grocery bag
- One plastic grocery bag
- One standard, black plastic trash bag (such as Hefty brand)
- One plastic trash bag that's marketed as biodegradable

- Four "twisty ties"

- Scissors

- Black plastic with which to line the "landfill"

- Shovel

- One large piece of wood or several smaller pieces with which to cover the "land-fill"

- Several bricks or good-sized stones

- Rubber or latex gloves

Suggested solid waste for each bag includes the following:

- One slice of bread

- Peels from four slices of apple

- 2 tablespoons coffee grounds

- One slice of cheese

- One paper towel, crumbled up

- Watermelon rind or banana peel

Just make sure you put the same amounts of the same waste in each bag. It's probably not a good idea to use meat or fish, which might attract more animals than the materials listed above.

# Conducting Your Experiment

This experiment is not difficult to perform, but it requires some preparation, and some adult supervision. You'll need to dig a fairly large hole in which to bury the four bags you'll be testing. The hole needs to be about a foot deep and about five feet wide.

**Standard Procedure**

Make sure you get permission from a parent to do this experiment before you choose this topic as your science fair project. You'll need your parents' cooperation in order to be able to do this project.

Once you've dug the trench, you'll need to line it with black plastic. You can buy black plastic in a roll, or you can simply use large-size trash bags. Lining the trench simulates a landfill. Landfills are required by law to have heavy liners in an attempt to prevent trash residue from leaching out into the ground and contaminating earth and water sources.

Once you've lined the trench, you'll continue your experiment as outlined here.

1. Cut the four trash bags so that they're all the same height.

2. Place identical solid waste ingredients into each bag, and secure the bags with twisty ties.

3. Place the bags side by side in the trench, then cover with dirt so that just the tops of the bags are showing.

4. Evenly water the soil over the bags.

5. Cover the trench with the boards, and weight them with some stones or bricks. This will help to keep any critters who smell the trash from digging up your experiment.

6. Wait for four weeks and uncover the bags. Remove enough dirt so that you can observe the bags, and record your observations. Remember to use your sense of smell, and be sure to wear rubber or latex gloves when handling the trash.

7. Replace the dirt over the bags, and put the wood back over top of the trench.

8. Four weeks later, uncover and dig up the bags again, making sure to note your observations.

9. Dispose properly of all trash, and fill in the trench so it's level with the ground around it.

If your mom or dad was a good sport and let you dig up part of the yard in order to conduct this experiment, be sure you restore the yard—or at least help to restore it— as closely as possible to its original condition.

# Keeping Track of Your Experiment

It's important in this experiment that you keep careful and detailed notes about what you observe. You can keep track of your observations in your journal, or you can make some simple charts to help you keep track of what occurs.

Your observations should include more than merely what you see when you dig up the trash bags. There probably will be odors you'll need to note, as well. Because four weeks will pass between the times that you observe the bags, you should notice some interesting changes in their conditions.

Be sure to begin your observations by noting the original conditions of the bags. Photographs of the bags at the beginning, middle, and end of the experiment would be extremely useful, and would enhance your display.

# Putting It All Together

Using your written descriptions and photographs, you'll be able to clearly demonstrate what happened to each trash bag while buried under the ground.

If you wanted to, you could cut a piece from each bag and include it as part of your display. You should, however, cover the pieces with a sheet of clear plastic wrap in order to prevent anyone from touching them.

# Further Investigation

There are many ways in which you could vary this experiment if you want to. One way would be to test the rate at which different types of paper bags degrade. Perhaps the brown grocery store bags break down faster than the shiny, colored shopping bags you get from American Eagle or your local department store. How does the paper sack in which you carry your lunch to school measure up?

Another variation would be to see if certain substances hasten the breakdown process of paper or plastic. You could experiment with lime (frequently used for home gardening), an acidic liquid such as lemon juice, hot water, and so forth. Just be sure to check with a parent about what you're allowed to use, and don't use anything that could be hazardous.

# Other Great Earth Science Projects

If you enjoyed doing this project, or even just found the idea of it interesting, you might want to consider another project in the *earth science* category.

Earth science is an extremely important area, and will become increasingly important as we face many challenges resulting from overpopulation, overuse of resources, weather changes, and so forth.

Included in this section are ideas for two other projects that fall under the category of earth science.

## Is One Room of Your House Colder Than Another?

You may have noticed in your house—especially if it has a basement or attic—that the temperature varies from one area to another.

In this suggested project, you'll attempt to identify the temperature variations of different parts of your house, and then to figure out what causes the differences to occur.

You'll use the temperature of a main living area as your control, and the variables will be other parts of the house. You don't need to record temperatures in every room, but pay special attention to areas on different levels (underground basements normally are relatively cool spots, while top-floor attics generally are warm), or rooms that might be built over the garage or other chilly location.

In order to conduct the experiment, you'll need to take daily air temperature readings both inside and outside your house. The outside temperature is important because it could cause the indoor temperature to vary. You'll also need to note special conditions like high winds and rain.

Record temperatures over a two-week period, and be sure to take them at the same times each day. You should check temperatures at least twice, and preferably three times a day.

**Basic Elements**

Earth science is simply the general term for sciences such as geology, geography, and geomorphology—all of which are concerned with how the earth is structured, how long it's been in existence, effects on the earth from various influences, and so forth.

**Standard Procedure**

Checking temperatures in several areas of your house two or three times a day for two weeks will result in a lot of numbers. Be sure to record your results carefully and accurately.

## What Factors Are Conducive to Fog?

If you've wondered about fog, and how and why it occurs, perhaps you'd be interested in a science fair project that explores when and how fog develops.

Basically, fog is just a cloud that hangs very low to the ground. When the temperature drops below the *dew point*—that's the temperature at which air becomes saturated with water vapor, and can't hold any more—the vapor condenses out of the moist air near the ground and forms fog. This usually occurs when warm air moves into a region over ground that is much colder. There are various types of fog that form for various reasons.

**Basic Elements**

The **dew point** is the temperature at which air becomes saturated with water vapor.

A suggested science fair project dealing with fog would be to keep a weather journal over the course of two months. It's best to do this in a season when the temperature is changeable, such as in the spring and fall.

Record the temperature at the same time each night and each morning, note the presence or absence of fog, whether skies are clear or cloudy, the speed and direction of the wind, and other meteorological factors.

After you've summarized and charted your observations, you should see some clear patterns of when fog occurred. Additional research will be necessary to help you figure out why fog occurs under certain conditions.

## The Least You Need to Know

- ◆ It's important that we find ways to reduce the amount of trash we generate, as well as materials that will degrade more quickly in landfills.

- ◆ This experiment requires two months to conduct, so be sure to begin the project well in advance of its due date.

- ◆ Because you'll be recording observations over a long period of time, you'll need to take thorough notes so you can remember what occurred. Photos would be very useful for this project.

- ◆ Earth science is an area of increasing importance as the earth faces a variety of challenges.

# Does It Matter How Much Air Is in Your Basketball (and Other Great Physical Science Projects)?

## In This Chapter

◆ Understanding air pressure

◆ Measuring the force of air pressure

◆ Air pressure is important in many ways

◆ Two sets of eyes and hands are better than one for this experiment

◆ Keeping your numbers straight

◆ The importance of physical science

Did you ever notice that some balls bounce better than others? Baseballs, for instance, don't have much bounce at all, except for when they land hard on the ground in front of you and jump up to nail your shinbone.

Footballs bounce, but you can never tell where they're going. Tennis balls bounce well, as do soccer balls and basketballs. But did you ever notice that some of the same types of balls bounce better than others?

Having the right ball for the game you're playing is important. Having a ball that does what it's supposed to do is equally important. In this chapter, we'll explore why some balls bounce better than others, and what you can do to assure that the ball you're using is in top condition for the game.

# So What Seems to Be the Problem?

If you have a ball that doesn't bounce properly, the immediate problem is that it will affect your game. The problem we'll attempt to solve during the course of this experiment, however, deals with air pressure.

**Basic Elements**

Air pressure, simply put, is the push that air has against all surfaces that it touches. The more air there is in a contained area, the greater the air pressure.

Specifically, we'll try to determine what amount of air pressure in a basketball makes the ball (and hopefully the person using the ball) perform best.

In the next few sections of this chapter, you'll learn a little bit about what *air pressure* is, and some of the effects it has on us and on our surroundings.

Air—as in the stuff we breathe—is matter. It has mass, or weight, and volume, meaning that it takes up space. It's funny to think about air as having mass and volume. You can't see it, and you don't sense that you're running into it as you run or walk. It's easier to get a sense of air and air pressure when you think of air that is contained within something, such as your basketball, an air mattress, or a bike tire.

Air is composed of different gases, including nitrogen, carbon dioxide, water vapor, oxygen, and others. All of these gases are composed of particles, or more scientifically, molecules, and they're all in constant motion.

As the molecules move about, they come in contact with surfaces of objects. The molecules push and press on those surfaces, exerting pressure on them.

It's that pressure that allows your basketball to keep its round shape and remain hard and bouncy. If air escapes from the ball, the pressure inside the ball changes. The problem you'll attempt to solve while doing this experiment is how these changes in pressure affect the ball.

If you like the title of this chapter, "Does It Matter How Much Air Is In Your Basketball?" you could use it as the title for your project. Or if you prefer, you could use one of these titles:

- ◆ Getting the Bounce You Need from Your Basketball
- ◆ How Much Pressure Is Needed for the Best Bounce?
- ◆ How Air Pressure Affects the Bounce of Your Basketball

# What's the Point?

Basketballs, along with other objects such as tires and footballs, are high-pressure products. This means that they require a lot of the force created by air pressure in order to get and stay fully inflated.

In the United States, we measure this force using psi, or pounds per square inch. Conventional wisdom is that a basketball needs 8 psi to be properly inflated. A car tire, just for a comparison, needs 40 psi.

Most basketballs have inflation instructions printed right on them. Take a look and you'll probably see something like "INFLATE 7 to 9 LBS." For this experiment, a ball inflated to eight pounds per square inch will serve as the control. The same ball, but containing different amounts of air, will provide the variables.

| Scientific Surprise |
|---|
| There are various measurements used for air pressure. Among them are pounds per square inch, millimeters of mercury, atmospheres, and pascals. |

Your challenge will be to determine if and how the amount of air inside your basketball causes it to bounce differently. You'll also determine whether the manufacturer's recommendation for the proper inflation rate gives the ball its best bounce.

# What Do You Think Will Happen?

If you've ever pumped up a bike tire, an air mattress, or a ball, you've had some experience with air pressure. Blowing up a balloon, which nearly everyone has done, is another example of air pressure.

Excessive contained air can cause such force against an object that the object bursts. This happens frequently with balloons, which are fairly fragile. It happens less frequently with basketballs, which are sturdier objects, but it can occur.

A loss of air pressure within an object such as a ball or tire, however, also can cause problems.

---

**Scientific Surprise**

Air pressure affects not only bouncing balls, but also people, animals, and the weather. If you've ever flown in a plane, you may have experienced some weird things going on with your ears due to changing pressure. Clear, sunny weather generally accompanies high-pressure systems, while low pressure can result in thunderstorms and other weather disturbances.

---

Think about observations you may have made that relate to air pressure. Does your bike ride as well when one of the tires (or maybe both) has lost air? Does your ball bounce as well? If you've ever played tennis with a ball that's lost pressure, you'll understand why they're referred to as "dead" balls. A dead ball hits the court and drops, as if it is—indeed—dead.

Using your experience and observations, come up with a hypothesis concerning your basketball and whether the amount of air in it will affect the way it bounces. Do you think that increasing the level of air even above the manufacturer's instructions will cause it to bounce better? What about if there's less air in it than recommended?

# Materials You'll Need for This Project

You need only three things—with two optional, additional items—to conduct the experiment for this project. They are:

- ◆ One basketball
- ◆ One air pump with a pressure gauge
- ◆ One tape measure or meter stick
- ◆ One long piece of paper and tape (optional)

It's a good idea to find someone to help you with this experiment, which involves dropping a ball and measuring how high it bounces. It would be possible—but very tricky—to do the experiment by yourself. Besides, basketball is always more fun with a friend!

# Conducting Your Experiment

To begin the experiment, you'll test how high the basketball bounces when it's inflated to 8 pounds psi—the recommended level for most models. From there, you'll increase or decrease the air pressure by putting more air into the ball or removing some air.

The trick will be to add or subtract air in measured increments so that you can keep track of where the pressure is.

Drop the ball close to a wall, so you can calculate on the wall the height of the bounce. It would be a good idea to make a paper measuring tape and hang it on the wall so you can better judge the height of the bounce. If you don't, be sure to watch closely how high the ball bounces and use your tape measure or meter stick to measure up to the spot on the wall at which the top of the ball stopped.

**Explosion Ahead**

Be sure to drop the ball each time from exactly the same height. Starting from different heights will make the results of your experiment invalid.

Make sure that your basketball is at 8 psi before you begin the experiment. Be sure to record both the pressure of the ball and the height of the bounce as you move through the following steps. You can record all of your results on the data chart found in the next section, "Keeping Track of Your Experiment."

Follow these steps to conduct your experiment:

1. Drop a basketball inflated to 8 psi from a height of 6 feet (1.8 meters). Be sure that the bottom of the ball—not the top—is at the 6-feet line when you drop it.

2. Observe, mark, and measure the height of the ball's bounce, remembering to measure at the top of the ball. Record your measurement.

3. Repeat steps one and two twice, recording your results as you go.

4. Using the air pump with the pressure gauge, increase the psi of the ball by one inch, to 9 psi, and repeat steps one through three.

5. Using the needle of the air pump, force as much air out of the ball as possible, then repeat steps one through three.

6. Increase the pressure within the ball to 1 psi, then repeat steps one through three.

7. Increase the pressure within the ball to 2 psi, then repeat steps one through three.

8. Increase the pressure within the ball to 3 psi, then repeat steps one through three.

9. Increase the pressure within the ball to 4 psi, then repeat steps one through three.

10. Increase the pressure within the ball to 5 psi, then repeat steps one through three.

11. Increase the pressure within the ball to 6 psi, then repeat steps one through three.

12. Increase the pressure within the ball to 7 psi, then repeat steps one through three.

Once you've completed all the steps, you'll need to average the height of the bounce recorded in each of the three trials for each step. To find the average, add the three measurements together, then divide the total by three.

*Recording bounce height.*

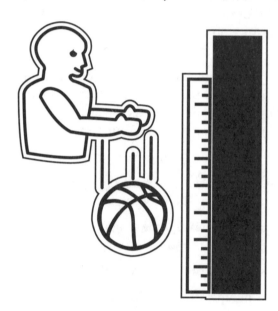

# Keeping Track of Your Experiment

As you conduct three trials for each of the basketball psi settings, you can record your information on the following data chart. Or you can make your own chart, if you wish.

# Putting It All Together

Once you've recorded all your data onto the chart, you can plot it onto a graph, letting the x axis represent the pressure, and the y axis represent the height to which the ball bounced when dropped from 1.8 meters, or 6 feet.

If you need a refresher course on graphs, go back and read the part of Chapter 6 that explains charts and graphs. Once you've completed your graph, you'll be able to clearly see if your findings prove, or disprove, your hypothesis.

**Basketball Bounce Height with Varying Air Pressure**

**Ball Dropped from 1.8 Meters (6 feet)**

| Air Pressure lbs/in$^2$ | Height Ball Bounced | | | | | | | |
|---|---|---|---|---|---|---|---|---|
| | Trial 1 | | Trial 2 | | Trial 3 | | Average | |
| | inches | cm | inches | cm | inches | cm | inches | cm |
| 8 | | | | | | | | |
| 9 | | | | | | | | |
| 1 | | | | | | | | |
| 2 | | | | | | | | |
| 3 | | | | | | | | |
| 4 | | | | | | | | |
| 5 | | | | | | | | |
| 6 | | | | | | | | |
| 7 | | | | | | | | |

*Use this chart to keep track of how high your basketball bounces at different levels of inflation. Be sure to drop the ball from a height of 1.8 meters (6 feet).*

# Further Investigation

If you're a hands-on kind of person, and you enjoyed doing this experiment, you can think about taking it a step or two further.

You could do this by testing the ball in the same manner, but varying the temperature conditions. You might try bouncing the ball outside on a very cold day, and then, using the same psi settings, record the height of bounces in a nice, warm house.

You also could test balls made of different materials, but inflated to the same level of pressure. Or you can come up with your own ideas for similar experiments.

# Other Great Physical Science Projects

Physical Science is a huge arena, encompassing a great deal of science. It deals with chemistry-related topics such as compounds, molecules, and the chemical elements. It also encompasses physics, electronics, and electricity.

Physical science can be as simple as air pressure within a basketball, or as complex as quantum physics. The laws of motion fall under the umbrella of physical science, and the first of two additional physical science projects you might consider deals with motion.

## Why Do Some Objects Fall Faster Than Others?

Gravity is a major player in the study of physical science. It is, of course, the force of gravity that causes objects to fall. One object always exerts a force of attraction on another object. This force of attraction is a pull, like the pull of gravity.

The larger an object is, the greater is the force of its attraction. Consider the fact that the sun, which is much, much larger than the earth, can, even at 90 million miles away, hold the earth and the other eight planets in orbit.

The moon, on the other hand, is much smaller than the earth, and has only about one-sixth of the gravity of the earth.

It was Italian scientist Galileo Galilei who formulated the laws of accelerated motion and free-falling objects. He found that when an object is dropped and falls to the ground it has a falling rate of 9.8 meters per second, squared.

You may wonder, then, why feathers float gently in the breeze instead of falling to the ground quickly, like a brick does.

> **Scientific Surprise**
>
> Astronauts who have walked on the moon feel light and weightless because there is very little gravity holding them down. On the other hand, if they were to go to Jupiter, which has much more gravity than the earth, they wouldn't even be able to lift a foot off the ground.

Well, it's because the air offers much greater resistance to the falling motion of the feather than it does to the brick. The air is actually an upward force of friction, acting against gravity and slowing down the rate at which the feather falls.

The brick, on the other hand, can cut right through the air as if it didn't exist. Galileo discovered that objects that are more dense, or have more mass, fall at a faster rate than less dense objects, due to this air resistance.

If a feather and a brick were dropped together in a vacuum—that is, an area from which all air has been removed—they would fall at the same rate, and hit the ground at the same time.

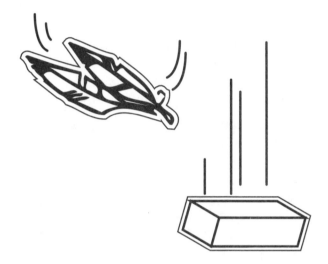

*A feather and brick dropped together. Air resistance causes the feather to fall more slowly.*

Understanding these basic facts will help you to be able to answer the question of why some objects fall faster than others.

You can test the rate at which various objects fall, noting both the mass of each object, and how long it takes for it to fall. Be sure to drop all objects from the same height, and be careful to use only objects that can't break. Record all your information in a journal, and chart your results.

Conduct three trials for each object so that you can calculate an average time.

## Do Objects Float Better in Salt Water Than in Fresh Water?

If you've ever floated in the ocean, you may have noticed that it's much easier to do so than it is to keep afloat in a swimming pool.

If you had to guess why that's the case, what would you say? The answer is just one word—salt.

When salt is dissolved in water, as it is in ocean water, that dissolved salt adds to the mass of the water and makes the water *denser* than it would be without salt.

Because objects float better on a dense surface, they float better on salt water than fresh water. The denser the salt water, the easier it is for objects to float on top of it.

You could make a science fair project out of this concept by measuring different amounts of salt into a specific amount of water and testing how well different objects float.

**Basic Elements**

**Density** is defined as mass per unit volume. If two objects are the same size, say a paper plate and a ceramic plate, we say that the ceramic plate is denser than the paper plate. The ceramic plate has the same volume, but much more mass.

A suggested method is to use five containers that are all the same size and shape. Put the same amount of water into each container. Use the first container as your control, and do not add any salt to it. Add 1 teaspoon of salt to the second container, two teaspoons to the third container, and so on.

Locate some objects that barely float in water, such as a paper clip, a small plastic ball, and a pen. Place the objects, one at a time, in the first container and observe how long they float in the water. Dry off each object and place it into the other containers in the same manner, observing carefully how long they remain afloat in the water.

Run three trials for each object in each container, recording all your information carefully and then graphing it.

There are hundreds of science fair projects dealing with physical science topics. Use your imagination to try to think of others you may enjoy doing.

## The Least You Need to Know

- Air pressure is the force of the molecules that make up air pressing against the surfaces they encounter.

- In the United States, we measure the force created by air pressure in pounds per square inch, or psi.

- Air pressure is important in getting a ball to perform at its best, but also affects the atmosphere and weather.

- You'll have a much better chance at completing this project successfully if you have someone to help you with the experiment.

- Because you'll be conducting three trials at each psi level, you'll need to work carefully to keep track of your numbers.

- Physical science is a huge area of science, encompassing many subjects, ideas, and topics.

# Part

# Intermediate-Level Science Projects

If you're looking at Chapters 12 through 16, chances are that you've already got a science fair or two or three under your belt. You know how they work, how the judging occurs, and so forth.

All that's left for you to do, then, is choose your topic and get busy. Just remember to choose a topic that is not only interesting to you, but seems manageable. Think about possible time constraints you might be facing between now and the project due date.

Some of the experiments in this chapter, particularly the one in Chapter 13, require significant time. You could always put that one on hold for another fair if you don't have time to complete it for the upcoming event.

# How Fast Are You? Testing and Measuring Reflexes (and Other Great Biology Projects)

## In This Chapter

- ◆ Understanding reflexes and reaction times
- ◆ The relevance of reaction time
- ◆ Identifying a control group
- ◆ Good news and bad news about the experiment
- ◆ Making your results as reliable as possible
- ◆ Understanding what's involved with biology

Think about the members of your family, and the families of your friends. Not just your immediate family, but your extended family. You may have grandparents—maybe even great-grandparents—and aunts and uncles, parents, cousins, and brothers and sisters.

As you no doubt have noticed, there are bound to be lots of differences among the members of an extended family. Some are old, others are young. Some might be tall, and others short. You might have cousins who have very blonde hair, while your and your sister's hair is brown.

Additionally, when you observe members of your family, or of other people's family, you'll notice that some move more quickly than others. And, chances are, it's the younger folks that are running laps around the older ones.

The gradual slowing down that normally occurs as a person ages doesn't mean that he or she gets lazy, or isn't able to be active. There are plenty of people in their 70s, 80s, and even 90s who exercise regularly. Some even participate in long-distance walking, biking, swimming, and running competitions.

Generally, however, older folks tend to take a little longer to do something that involves physical movement than younger ones. Their bodies slow down, it takes longer to complete tasks, and they experience a slowing in reaction time, as well. In this chapter, we'll look at the issue of reaction time—or reflexes—and how it varies between younger and older people.

# So What Seems to Be the Problem?

The problem that we'll attempt to solve with the experiment outlined in this chapter is whether older people have slower reaction times than younger people.

From the content of this chapter so far, and from your daily observations, you probably have an idea that indeed, that is the case.

Even so, it's a question worth investigating, and this experiment will enable you to do so. First, however, let's take a look at exactly what reaction times and reflexes are.

A person's *reaction time* is a measure of how fast they can respond to a situation or stimulus. Reaction times are linked to *reflexive actions*, or actions that you take without first thinking about them.

This science fair experiment uses an electronic toy to measure the reaction time of people of different ages. It's easily done, but requires careful record keeping so that you'll be able to track and record results.

**Basic Elements**

Reaction time is a measure of how fast a person can respond to a situation or stimuli. A **reflexive action** is an action performed automatically, without any advance thought.

You could use "Does Age Affect Reaction Time?" as the title for your science fair project. Or, you could choose from one of these titles:

♦ Testing Reaction Times of Various Age Groups

♦ Is Reaction Time Age Related?

♦ Reaction Time: Kids vs. Adults

Keep reading to see how the problem you'll be attempting to solve during the course of this experiment relates to everyday life.

# What's the Point?

If you're driving or riding your bike, and the person in front of you stops without any warning, your reflexive action probably will be to slam on your brakes, or to swerve out of the path of the car or bike before you.

Whether you manage a safe stop or end up hitting the vehicle in front of you depends largely on your reaction time. The quicker you get the brakes on or begin steering away from the other vehicle, the better chance you have at avoiding an accident.

It's important to have an idea of how good your reaction time is, and to keep an eye on it as you get older. Studies have shown that, typically, a person's reaction time is at its best when a person is in his or her late 20s. Reaction time improves from infancy into the late 20s, due, it's thought, to increased experience.

| Scientific Surprise |
|---|
| You probably know that alcohol and other drugs can adversely affect a person's reaction time. But do you know that other things, including lack of sleep and even some foods that you eat, can affect reaction time, as well? |

For example, you'll have a better idea how to react, and presumably will react more quickly, the fourth or fifth time that a car or bike stops suddenly in front of you than the first time it happens.

After the late 20s, however, reaction times typically begin slowing down. The slow-down occurs gradually until the 50s or 60s, and gets more pronounced as a person gets into his or her 70s and beyond.

The experiment in this chapter will allow you test these statements, and find out whether the older folks in your family have reaction times that are typical of their ages, or whether they defy scientific findings.

The control you'll use in your experiment will be a group of people who are between the ages of 20 and 30. This group should be at its reflexive peak. Other groups of people—both older and younger—will represent the variables.

Of course, reaction time will vary from person to person, and you can't assume that every 26-year-old will react more quickly than every 18-year-old. Plus, the difference in reaction times between people of various ages may be tiny—measured in increments of seconds, not seconds. Still, you should be able to gather enough information to make a comparison.

# What Do You Think Will Happen?

Based on what you've read and observed, chances are that your hypothesis is that younger people react more quickly than older ones. If you've reached that conclusion, then you can carefully conduct your experiment to learn if your results will support your hypothesis.

During the experiment, you'll test the reflexes of people in different age groups using an electronic toy available in toy stores and from Internet sources. The toy is most commonly known as a Bop It Extreme.

Before we tell you exactly how the experiment will work, take a few minutes to think about what you've observed while watching people over the course of your lifetime. Does it seem to you that younger folks are quicker and react faster than older ones? Do you think that someone in his or her 20s has a better reaction time than you do?

So go ahead and take your best guess, and then begin preparing to do the experiment that will test your hypothesis.

# Materials You'll Need for This Project

You'll need to have a Bop It Extreme for this experiment. You should be able to find one in your local toy store, or you can order one on the Internet.

The Bop It Extreme gives random, rapid commands to which the user must respond. It forces the person playing to respond quickly, before the next command is issued. The user is told to flick, spin, pull, or twist one of four extending parts that correspond to the commands. One point is scored each time a player performs a command correctly within the short, allotted time period.

*A Bop It Extreme will allow you measure the reflexes of people of various ages.*

The turn ends when the player is no longer able to complete the tasks of flicking, pulling, spinning, and twisting. The machine notes each action performed correctly with a drumbeat sound, and every 10 correctly completed actions with a crashing cymbal sound.

Once you have the Bop It Extreme, the only other materials you'll need for this experiment are as many people of different ages as you can round up, and paper and pencils with which to record your results.

**Standard Procedure**

You can get a Bop It Extreme from Amazon.com for $19.95. Just go to www.amazon.com and type "Bop It Extreme" in the search box. You also can find the toy on the eBay site.

## The Procedure

You'll need to test a lot of people in order to complete this experiment, and that will require some time and planning. The good news is that you can space it out over a period of time, just as long as you begin far enough in advance.

**Standard Procedure**

We've included age groups from 0 to 9 through 90 to 99. If you don't know people in their 80s or 90s who are willing or able to participate, you can skip those groups.

Plan to take the Bop It Extreme with you to family gatherings, your friends' houses, and other places so that you get a chance to test people of different ages. If you know someone who lives in an apartment or other facility for elderly people, you may be able to test all of your older participants in just one visit. Remember to take your notebook or journal along so you can record scores.

You'll need to have at least three people from each of the age groups listed here:

- ◆ 0 to 9
- ◆ 10 to 19
- ◆ 20 to 29
- ◆ 30 to 39
- ◆ 40 to 49
- ◆ 50 to 59

- ◆ 60 to 69
- ◆ 70 to 79
- ◆ 80 to 89
- ◆ 90 to 99

**Explosion Ahead**

Try to avoid including people in your experiment who are overly familiar with using a Bop It Extreme. Familiarity with the machine may enable someone to perform better than they would ordinarily.

Five people from each age group would be better than three, but if you can't locate that many, three will do. Just be sure that you have the same number of people from each category.

*Using the Bop It Extreme, compare the reflexes of a young person with those of an older one.*

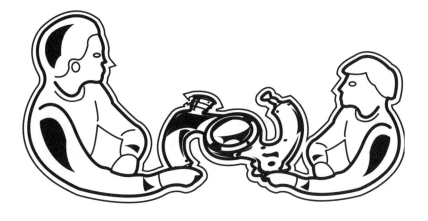

Once you've identified your participants, you should proceed with your experiment, following the steps listed below. This procedure will be the same for each person participating.

1. Prepare a chart on which to record information from each person who participates. Your chart should look like the one shown in the next section, "Keeping Track of Your Experiment." If you want, you can record the information directly onto that chart, but you'll need to recopy it if you want to include the results with your display.

2. Give a player three practice turns so he or she is familiar with how the Bop It Extreme operates.

3. Have a player take three turns (three trials) and record the number of actions correctly completed.

4. Calculate the average number of points scored to the nearest whole number and record that number on the data chart.

5. Repeat these steps for the rest of the people you'll be testing. Ideally, you'll test 45 people, or 5 from each group.

# Keeping Track of Your Experiment

This experiment requires that you keep good records, since you'll be working with so many people. Be sure to record your data accurately as each person participates.

This first chart is the one on which you'll record information about each participant.

On the second chart, you'll record the average scores of each age group. Once you've averaged the scores of each group, you can make a bar graph illustrating that information.

Once you've averaged the scores of each age group, it shouldn't be difficult to use that information to make a bar graph. See Chapter 6 for more information about charts and graphs.

### Standard Procedure

To find the average score of each age group, add together the scores of all the people within the group, then divide that number by the number of people in the group. That number will be the average score.

| Name | Age | Score | | | |
|---|---|---|---|---|---|
| | | Trial 1 | Trial 2 | Trial 3 | Average |
| 1 | | | | | |
| 2 | | | | | |
| 3 | | | | | |
| 4 | | | | | |
| 5 | | | | | |
| 6 | | | | | |
| 7 | | | | | |
| 8 | | | | | |
| 9 | | | | | |
| 10 | | | | | |
| 11 | | | | | |
| 12 | | | | | |
| 13 | | | | | |
| 14 | | | | | |
| 15 | | | | | |

*Chart 1: Player Information.*

**Average Scores by Age Group**

| Age Group | # of People in This Age Group | Average Score |
|---|---|---|
| 0 to 10 | | |
| 11 to 20 | | |
| 21 to 30 | | |
| 31 to 40 | | |
| 41 to 50 | | |
| 51 to 60 | | |
| 61 to 70 | | |
| 71 to 80 | | |
| 81 to 90 | | |
| 91 to 100 | | |

*Chart 2: Average Scores by Age Group.*

# Further Investigation

If you enjoyed conducting this experiment, you could take it a step further by noting the time of day at which each participant played the game. Do you think that the time of day might affect someone's reaction time?

Because studies suggest that males typically have faster reaction times than women, you also could note the sex of each participant and see if that theory proves to be correct or incorrect.

# Other Great Biology Projects

Biology—commonly known as the science of life—encompasses many, many areas. Biology concerns itself with various aspects of plant life, animal life, and human life. It deals with the origin of life, its distribution, physiology, development, habits, and so forth.

If you're interested in the field of biology, or simply enjoy working with people, animals, or plants, you might want to consider doing one of the experiments explained below. Both of these projects fall under the category of biology.

## Determining Your Speediest Shoes

You probably have at least a couple of pairs of sneakers, or at least what used to be called sneakers. These days, specialized sneakers are known as running shoes, cross-trainers, basketball shoes, and so forth. In addition to your sneakers, you probably have shoes such as sandals or dress shoes that you wear for other occasions.

You can do an experiment to test which pair of shoes helps you to run the fastest. It's not hard to do, and you'll have fun.

To do this experiment, you'll need four pairs of shoes. Your control should be your favorite pair of running shoes. Even if they're not shoes designed especially for running, they should be the ones in which you feel most comfortable when you run.

The variables you'll use are as follows:

 ◆ Another pair athletic shoes—maybe those you wear to play soccer, softball, or basketball

 ◆ 1 pair sandals

 ◆ 1 pair dress shoes

### Standard Procedure

If you have a treadmill and prefer to run indoors, set your treadmill for a quarter or half mile and time how long it takes you to complete that distance wearing the different types of shoes.

If you don't have dress shoes or sandals, feel free to substitute a pair of boots or whatever other shoes you might have.

To conduct the experiment, choose a running area. This could be the track at your school, around your block, or in a vacant lot near your home. Wearing the four different pairs of shoes over four days, run the distance you've selected, recording your time with a stop watch, or having someone else time you.

You should do three trials for each pair of shoes, so don't select a running area that's two miles long unless you're well trained and up for a whole lot of running.

Doing the experiment over four days, preferably at the same time each day, assures that you'll be in the same condition for each run. If you were to do all the runs in one day, you'd become tired and your times would probably increase.

Record all your observations and data on neatly designed charts. You could even make a line graph, using a different colored pen to represent each shoe type. At the end of your experiment, relate how your hypothesis measured up to your conclusion.

## Does Caffeine Increase Heart Rate?

Does your mom or dad like to have a cup or two of coffee in the morning? Does your sister have a cola habit? If so, they can help you with this experiment, aimed at discovering whether *caffeine* affects the rate at which a heart beats.

The basics of your experiment would be to measure and record the resting heart rate of your subject. You'd then have that person drink a cup of coffee, some soda, or other beverage containing caffeine. You can determine the amount of caffeine they'll be ingesting by reading the label of the soda or coffee can.

Wait for 20 minutes after the beverage has been consumed, and then recheck the resting heart rate. Just be sure that your subject sits quietly during and after the time he or she is consuming the caffeine. Exercising would cause the heart rate to increase, with or without the benefit of caffeine.

**Basic Elements**

**Caffeine** is a substance found naturally in the leaves, fruits, and seeds of more than 60 kinds of plants. These include tea leaves, coffee, kola nuts, and cocoa beans. It's found in various foods and drinks that are commonly consumed, including coffee, tea, chocolate, and some soft drinks. It also can be added to foods in which it does not naturally occur.

Ideally, you could do this with three or four different people, and conduct three trials for each person. Record your results, and see whether the caffeine affected their heart rates.

## The Least You Need to Know

◆ Reflexes and reaction times are closely related in that the better your reflexes are, the faster your reaction time will be.

◆ Reaction time is important in nearly every area of human life.

◆ Studies have shown that reaction time normally peaks in the late 20s, then gradually decreases.

◆ This experiment requires you to round up a lot of people of all different ages, which may make it difficult for some.

◆ Various factors can affect reaction time, so it would be best to conduct the experiment three times on each person to assure more reliable results.

◆ Biology is often called life science, and deals with the origin, habits, distribution, and other aspects of living things.

# Does the pH of Water Affect the Growth of Bean Plants (and Other Great Botany Projects)?

## In This Chapter

◆ Understanding acids and bases

◆ The effects of positive and negative ions

◆ Different soils for different plants

◆ Finding the materials you'll need

◆ Guarding against contamination

◆ Considering other botany projects

Although this chapter is classified as a botany project, you'll notice as you work through it that it contains a fair amount of chemistry, as well.

It's not unusual for scientific areas, or disciplines, to cross over in the course of a project or experiments. This project is a good example of how that occurs.

You've probably heard of the pH scale, or heard someone talk about the pH factor of a particular material. But what is pH exactly, and how does it affect the growth of plants?

In this chapter, we'll explore the basics of pH, and experiment to learn how the pH factor of liquid affects the germination and growth of bean seeds. By the time you finish, you'll have had valuable lessons in both botany and chemistry, and have a better understanding of how branches of science overlap.

# So What Seems to Be the Problem?

You know that plants need certain things to help them grow. They need some kind of growing medium, usually dirt. They need light, and they need water.

The problem you'll attempt to solve while doing this science fair project is whether the pH of the water with which plants are sprinkled affects the rate of growth.

To get a better idea of what you'll be doing, and to help you formulate a hypothesis, it's important that you have a general understanding of exactly what pH is.

The initials pH stand for percent hydronium ion. The pH scale is used as a measure of how acidic or basic a liquid is. But how do liquids become acidic or basic? Isn't a liquid just a liquid?

**Basic Elements**

An **ion** results from the loss or gain of one or more electrons from an atom, causing either positive or negative ions to form.

Water—and distilled water, at that—is the only liquid that is neutral. That means it's right in the middle of being acidic or basic—and it's neither. It's just pure water.

The pH scale starts at zero and ends at 14. The more acidic a liquid is, the lower its number on the pH scale. The less acidic—or more basic a liquid is—the higher its number.

Most of the liquids you encounter on a daily basis are just around neutral. They might be a little above or a little below, but most liquids tend to be closer to neutral than at either end of the pH scale.

Liquids get their pH level as a result of molecules that split apart to form positive and negative ions. An *ion* is the loss or gain of electrons from an atom. When an atom loses electrons, it forms a positive ion. When an atom gains electrons, it forms a negative ion.

Liquids will be either acidic or basic (also called alkaline), depending on whether they contain positive or negative ions. If there are more positive ions in the water, the water is more acidic. If there are more negative ions in the water, the liquid is more basic.

In this experiment, you'll control the pH of the water you'll use on bean plants by adding certain substances to make distilled water either acidic or basic. You'll also control all other factors, such as how much water and light each plant gets.

| Scientific Surprise |
| --- |
| A negative ion, formed when an atom gains electrons, is also called an anion. A positive ion, formed when an atom loses electrons, is also known as a cation. |

If you want to, you can use the name of this chapter, "Does the pH of Water Affect the Growth of Bean Plants?" as the title for your project. Other names to consider might be:

- What Type of Soil Do Bean Plants Prefer?
- Acid or Alkaline—What's Right for Bean Plants?
- To Grow or Not to Grow: Acidic vs. Alkaline Soil for Bean Plants

When you've finished with the experiment, you'll know whether bean plants prefer water that is acidic or basic.

# What's the Point?

Some plants prefer acidic conditions. We call these acid-loving plants. Acid-loving plants include the following:

- Holly, pine, fir, spruce, birch, oak, magnolia, willow, and flowering crabapple trees
- Rhododendrons and azaleas
- Hydrangeas
- Roses
- Cranberry, strawberry and blueberry plants
- Mountain laurel
- Crocuses

Other plants, however, such as the ones listed here, prefer alkaline soil:

- Yew, boxwood, and barberry shrubs
- Flowering plum and cherry trees

- Ash, beech, filbert, and maple trees
- Clematis
- Mock orange
- Lilacs
- Pinks
- Asters

Gardeners often help plants along by making the soil in which they grow either more acidic or more alkaline. There are products available, such as Miracid, that boost the acidity of soil. Garden lime (its chemical name is calcium carbonate) will help make soil alkaline.

In the experiment described below, you'll use distilled water as your control, and water with varying pH levels as your variables. This will allow you to observe the effects that liquids of varying pH levels have on the bean plants.

Who knows? You may end up increasing your interest in, or developing an interest in, gardening through this project. If nothing else, it will give you a better understanding of how plants grow and what types of factors affect them.

# What Do You Think Will Happen?

If you've had experience with gardening and growing plants, you know that plants react differently to all sorts of factors.

Some plants like to be watered frequently, while others like to wait for a drink until the soil is completely dry. Some are much more susceptible to heat or cold than others. Some plants thrive in sunlight, while others like shady conditions. You've already read that some plants prefer an acidic soil, while others like basic soil.

You can do some research about growing bean plants to help you form a hypothesis. If you don't have much experience with plants, or don't have a good understanding of pH, it probably would be beneficial for you to learn more about growing plants, different types of soil, and so forth.

Or you can simply consider what you may already know about growing plants and make an educated guess about what will happen to the bean plants with which you'll be working.

# Materials You'll Need for This Project

One thing about this experiment is that it's going to take some advance planning and a significant amount of time. You'll need almost a month from the time you plant the seeds until the time you draw final conclusions about the growth of the plants.

You also will need a material or two with which you may not be familiar. Most of the materials you'll need though, are common household items. You'll need:

- ◆ A substance used to adjust the pH level of water. We suggest a set of products called pH Up and pH Down, a brand that's readily available in pet supply stores. You'll need a bottle of each pH Up and pH Down. Retail cost is about $3.50 a bottle.

- ◆ pH test strips or test kit. You also can purchase these supplies from your local pet supply store, hobby shop, or from online sources. A package of test strips in a hobby shop should run you somewhere about $3 or $4. You should have 50 strips to make sure you have enough for the experiment. Or, your science teacher may have extra test strips that you could get for this project. It doesn't hurt to ask, right?

- ◆ Seven large-size plastic drinking cups. Cups should be about 16 ounces to allow room for bean plants to grow.

- ◆ Soil to fill the cups about three-quarters full. You'll need to use the same soil for all the cups. Buying a bag of potting soil is recommended.

- ◆ Twenty-one bean seeds

- ◆ Distilled water

- ◆ Metric ruler

- ◆ Seven two-liter, plastic bottles, empty and washed well

- ◆ Small paper cups in which to measure water

If you're going to order supplies from an online provider, be sure to do so ahead of time so you don't get stuck, unable to begin your project.

# Conducting Your Experiment

Because you need seven individual bottles of water and seven cups, you'll need some space to set up this experiment.

It's very important that the cups containing the bean seeds are all kept in the same conditions. They each need to have the same amount of light, heat, and so forth.

**Explosion Ahead** _____

Contaminating one bot-tle of water with any water from another bottle will affect the results of your experiment. Try as hard as you can to keep water with different pH levels com-pletely separated.

**Standard Procedure** _____

Pour the water into the soil, not on the leaves of the plants. Plants take up water from their roots, not their leaves.

And it's extremely important that each plant receives the same amount of water. If you give the plants different amounts of water, you'll be unable to determine whether the plant was affected by the pH of the water, or simply the varying amount.

You can't mix water from any of the two-liter bottles, or use the same container to hold water from different bottles without washing it out in between. That's why it's recommended that when you water plants, you use small paper cups, pouring water from each two-liter bottle into its own paper cup, and then onto the bean plant.

If your plants don't need a full cup of water, measure up to the halfway mark of each cup and make a line. Fill the cup with water up to the line to assure that each plant gets the same amount.

Follow these steps to conduct the experiment:

1. Starting with seven two-liter bottles of distilled water, prepare each bottle so it has a specific pH value. Leave one bottle untreated (your control), with a pH level of 7. Add pH Up or pH Down to the other bottles so that one bottle has a pH level of 1, one has a pH level of 3, one has a pH level of 5, one has a pH level of 7, one has a pH level of 9, one has a pH level of 11, one has a pH level of 13. You'll raise or lower the pH level for each bottle from 7, depending on whether you're making the water acidic or alkaline. Cap the bottles tightly, label each one so you know which is which, and place them in an undisturbed location.

**Explosion Ahead** _____

While it's interesting to see how pH level affects plants, don't be tempted to drink any of the water yourself, and avoid splashing it on your skin, eyes, and so forth. Some of the treated waters you'll be using have very high or low pH levels, and should only be used for the purposes of this experiment.

Do write down how many drops of either pH Up or pH Down you need to add to each two liter bottle. If you need to make more water at the different pH levels, you'll already know how much of the pH product to add, and the second round of water will have the exact same pH levels as the first round.

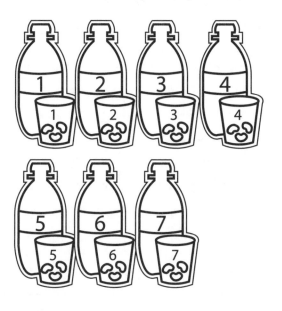

*Labeling each cup and each bottle will help you keep track of which water you'll use for each plant.*

2. Plant three bean seeds into each of the seven large, plastic cups that have been filled about three-quarters of the way with potting soil.

3. Mark each cup and match one cup with one two-liter bottle of water. It's extremely important that each cup is watered from the same bottle each time, and not from any other bottle.

4. Making sure each cup gets the same amount of water, water the seeds in each cup so that the soil is moist, but not saturated. You might want to transfer water from the two-liter bottle to the cup with a tablespoon, allowing you to exactly measure how much each pot will get. Just be sure to wash the tablespoon between using it to handle water from different bottles.

5. Observe the cups every day, watering when the soil appears dry. Just be sure to always give the seeds in each cup the same amount of water. The plants will require more water as they get larger.

6. Using a metric ruler, measure the plants and record your observations every four days. Record the growth of the plants in a chart like the one found in the next section, "Keeping Track of Your Experiment."

Your experiment will be finished after 28 days, meaning you will have measured each plant seven times.

# Keeping Track of Your Experiment

Make sure to note the height of each plant within the seven cups—21 plants in all—each time you measure. Keep track of how much water you've given the plants as well. The amount of water given to the plants in each cup should be the same.

Data charts to record water application and plant growth are provided, or you can make your own charts, if you prefer.

To calculate the average height for the three plants in each cup:

1. Add the three heights together for each of the three plants in cup #1.

2. Divide that total by 3.

3. Record this number in data chart 2.

Repeat steps 1–3 for the remaining six cups.

**Heights of Each of the Three Plants in Seven Different Cups (Measured in Centimeters)**

| Date Height Taken | Cup #1 pH 1 Plants | | | Cup #2 pH 3 Plants | | | Cup #3 pH 5 Plants | | | Cup #4 pH 7 Plants | | | Cup #5 pH 9 Plants | | | Cup #6 pH 11 Plants | | | Cup #7 pH 13 Plants | | |
|---|---|---|---|---|---|---|---|---|---|---|---|---|---|---|---|---|---|---|---|---|---|
| | 1 | 2 | 3 | 1 | 2 | 3 | 1 | 2 | 3 | 1 | 2 | 3 | 1 | 2 | 3 | 1 | 2 | 3 | 1 | 2 | 3 |
| | | | | | | | | | | | | | | | | | | | | | |
| | | | | | | | | | | | | | | | | | | | | | |
| | | | | | | | | | | | | | | | | | | | | | |
| | | | | | | | | | | | | | | | | | | | | | |
| | | | | | | | | | | | | | | | | | | | | | |
| | | | | | | | | | | | | | | | | | | | | | |
| | | | | | | | | | | | | | | | | | | | | | |

*Use this chart to record the height, in centimeters, of each plant in each of the seven cups.*

**Average Plant Height (Measured in Centimeters)**

| Date Height Taken | Cup #1 pH 1 | Cup #2 pH 3 | Cup #3 pH 5 | Cup #4 pH 7 | Cup #5 pH 9 | Cup #6 pH 11 | Cup #7 pH 13 |
|---|---|---|---|---|---|---|---|
| | | | | | | | |
| | | | | | | | |
| | | | | | | | |
| | | | | | | | |
| | | | | | | | |
| | | | | | | | |

*Use this chart to record the average height of the plants in each cup.*

Don't assume that all plants within a cup will grow equally. You can average the height of the three plants within each cup.

# Putting It All Together

Use the information contained on your data charts to make some graphs, plotting the growth of each plant. Or, you can summarize your observations in written form, if you prefer.

Did some of the plants grow quickly at first, only to halt their growth later? Did any of the plants die? Did your hypothesis prove to be correct?

Look for trends and patterns, concluding which plants had the best overall growth, the fastest starts, and so forth.

**Amount of Water Used to Water the Plants (Measured in Milliliters)**

| Date Plants Watered | Cup #1 pH 1 | Cup #2 pH 3 | Cup #3 pH 5 | Cup #4 pH 7 | Cup #5 pH 9 | Cup #6 pH 11 | Cup #7 pH 13 |
|---|---|---|---|---|---|---|---|
| | | | | | | | |
| | | | | | | | |
| | | | | | | | |
| | | | | | | | |
| | | | | | | | |
| | | | | | | | |
| | | | | | | | |

*Use this chart to record the amount of water given to each cup on a particular date.*

# Further Investigation

If you enjoyed this experiment and want to try another variation of it, you could try to grow plants hydroponically—that is, in water instead of dirt—while varying the pH level of the water.

It would be interesting to see if you get different results than you did when you used water of varying pH levels to water plants growing in dirt.

You also could put liquids other than water on plants to see how growth was affected. Or, you could work with acid-loving plants and base-loving plants, testing to see at what pH level they grow best.

# Other Great Botany Projects

Botany is a fascinating area of science, and one that is applicable to everyday life. Botanists work hard to improve plants—making them more resistant to drought and insects and higher yielding.

If you like working with plants and would like to try another experiment, there are two suggested below. One explores the process of altering plants by cross-pollinating different types of plants, and the other explores whether bean plants grow better in soil or in water.

## Creating New Plants Through Cross-Pollination

The process of cross-pollinating occurs all the time in nature. Pollen is exchanged from one flower to another via butterflies, bees, and wind. In addition, botanists work closely with cross-pollination as a means of altering plants.

You can try to find out what happens when you cross-pollinate—or remove pollen from one type of flower and put it onto another type. It's not difficult to do this, but you'll need to have a basic understanding about the male and female parts of a flower, and on which part of a flower the pollen must be placed in order to result in fertilization, and ultimately, a seed.

Flowers contain both male and female parts, and often pollinate themselves as pollen moves within the flower from the male part to the female part. You can, however, move pollen from one type of flower onto another, essentially creating a new type of plant.

**Standard Procedure**

To learn more about the genetics of plants, read about the work of Gregor Mendel, known as the father of genetics. Mendel, who lived from 1822 until 1884, conducted many experiments dealing with genetics and heredity. His most famous experiments were conducted using pea plants.

If you're interested in botany and this project, begin by researching the anatomy of flowers, and the processes of pollination and cross-pollination. You can use any flowers you want, but be sure to leave plenty of time, as you may have to repeat the experiment.

Once you put pollen from one flower on another, you'll have to hope that fertilization occurs. If not, you'll need to try again.

## Do Bean Plants Grow Better in Soil or in Water?

If you read Chapter 8, you already know a little bit about hydroponics.

Hydroponics is the practice of growing plants in water instead of soil. Many plants can be grown successfully in this manner. If you're interested in hydroponics, you might want to do an experiment in which you test whether bean plants (or another type of plant, if you prefer) grow better in soil or in water.

**Standard Procedure**

To learn more about the basics of hydroponics, check out www.vegsource.com, or www.sea-of-green.com.

You could start with seeds or plants of the same size, planting one in soil and keeping the other in water, while making sure other factors such as heat and light are the same.

Hopefully, you've gotten an idea of what an interesting area of science botany is, and will think about possibilities within the field for future science fair projects.

## The Least You Need to Know

- Liquids can be acidic, basic, or neutral, depending on whether they contain positive or negative ions.

- Some plants like acidic soil, while others like a basic, or alkaline, soil.

- The specialty materials you'll need for this experiment can be purchased in a pet supply store or over the Internet.

- It is extremely important to make sure the only variable in this experiment is the pH level of the various bottles of water.

- Watch carefully to make sure water with one pH level doesn't in any way get mixed with water that has a different level.

- Botany is a fascinating area of science within which you could come up with many science fair projects.

# Chapter 14

# Are All Pennies Created Equal (and Other Great Chemistry Projects)?

## In This Chapter

- ◆ A brief history of the penny
- ◆ Knowing what a penny is made from
- ◆ Using facts to formulate a hypothesis
- ◆ Walking you through the experiment
- ◆ Drawing conclusions and presenting results
- ◆ Thinking about other chemistry-based experiments

The penny, which has been around in the United States in one form or another since 1787, was the first currency of any type authorized by the newly formed America. Benjamin Franklin, who is well known for his famous quotation regarding the penny ("A penny saved is a penny earned"), suggested the first design for the new coin.

The original penny was 100 percent copper and was known as the Fugio cent. It was made at a privately owned mint. That particular model of penny lasted until 1859, when the Indian Cent was introduced. If you've never seen an Indian penny, be sure to check your change. They still show up now and then, and they're very cool. The Lincoln penny—our current model—first appeared in 1909 to mark the 100-year anniversary of Abe's birth.

The Lincoln penny has undergone numerous design changes as well as changes in composition. During World War II the composition of the penny was changed from 95 percent copper and 5 percent zinc and tin to zinc-coated steel because copper was needed for war efforts.

The problem you'll try to solve in this chapter, however, doesn't concern the design of pennies or what they will buy. Instead, you'll explore the chemical composition of pennies minted during the past 30 or 35 years, trying to figure out if and when it has changed. Let's get started.

# So What Seems to Be the Problem?

The problem, or question you'll attempt to answer during the course of this project, is whether the chemical composition of the penny has changed over the decades since 1970. In other words, are different materials used to make newer pennies than those that were used to make older pennies?

| Scientific Surprise |
| --- |
| The penny is the most circulated American coin. More than 300 billion one-cent coins have been produced since they first showed up in 1787. |

Pennies minted before 1970 look pretty much the same as those minted after. They're probably dirtier, but you can't tell by looking at them whether they're made of the same materials.

By the time you finish the project, however, you'll know whether the composition of those pennies you carry around in your purse or pockets has changed during the past 30 or 35 years.

If you want to give a name to your project, you could use the chapter title, "Are All Pennies Created Equal?" A few other suggestions for project titles are:

- How Have Our Pennies Changed?
- A Penny Is a Penny—or Is It?
- Exploring the Chemical Composition of a Penny

Now that you've identified the problem you'll be attempting to solve, it's time to consider the purpose of this project.

# What's the Point?

The point of this project is to determine, using the scientific method, whether pennies made before 1970 (those will be your control group) are heavier or lighter than those made in each decade after that date (the groups of pennies from each decade since the 1970s will be your variable groups). You also should be able to get an idea of specific years in which the weight of the penny was changed.

So why are we suggesting this topic for a science fair project? Who really cares what materials are used to make one-cent coins, anyway? It doesn't change the way they look or what you can buy with them, right?

Although there may be no real practical need to know the composition of our pennies, it's interesting to think about how they may have been changed and why. If you do determine that pennies made in the decades since 1970 are different from those minted before that, maybe you'll want to do some additional research and try to find out why. Is it because of a copper shortage, like it was during World War II? Or maybe there's a different reason.

You can't tell by looking at pennies whether they're made of exactly the same materials. The older ones are the same size as the newer ones, so they look the same. The only way to establish whether the *chemical composition* of the different groups of pennies is the same or different is to determine the *mass* (weight) of each group.

Different metals have different densities. A penny containing more of a certain metal than another penny will have a different mass, because its *density* is different.

**Basic Elements**

The **chemical composition** of an object is the materials of which it is made. The **mass** is its weight, and the **density** is the mass per unit volume, which is usually measured in grams per milliliter. The density of water, for instance, happens to be one gram per milliliter.

If the average mass of your control group (those pennies minted before 1970) is different from that of any of the variable groups, you'll know that the pennies are not made of the same amounts of the same materials.

# What Do You Think Will Happen?

Take a few minutes to think about the facts stated below, and then try to work out your hypothesis.

♦ A penny is made of copper and zinc.

♦ Copper is heavier than zinc.

♦ The cost of copper has been on the rise during the past couple of decades.

♦ It's less expensive to mint a penny that contains an increased amount of zinc.

> **Scientific Surprise**
>
> The mass of one cubic centimeter of copper is 8.96 grams, while the mass of the same amount of zinc is 7.13 grams.

Once you've considered these facts, you should be able to make an educated guess concerning the results of your experiment. But you won't know for sure until the experiment is completed.

# Materials You'll Need for This Project

There are very few materials needed for this experiment. The only things you'll need are listed below.

♦ An electronic or digital balance scale or other tool to measure mass.

♦ Enough pennies to produce 10 of each group. You'll need 10 pennies minted before 1970, 10 minted from 1970 through 1979, 10 minted between 1980 and 1989, 10 minted between 1990 and 1999, and 10 that have been minted since 2000.

> **Explosion Ahead**
>
> Don't be tempted to use your bathroom scale for this experiment. The weight difference between your penny groups may be very, very small, and your bathroom scale probably isn't sensitive enough to pick up the difference.

A traditional balance scale is a scale that has two pans that hang from opposite ends of an overhead arm. If objects placed in the different pans are of different weights, the pan holding the heavier object will be lower than the pan containing the lighter object. There are many electronic versions of balance scales available. You can purchase one in your local office supply store.

If you happen to have a balance scale in your home or can borrow one, that's great. If you don't have one, ask your science teacher if the school has one you can use. If you can't take it home with you, you can easily carry your pennies along to school and weigh them there.

If you don't have a piggy bank you can break into and rob, you can easily get pennies from your local bank. A roll of pennies contains 50 coins, so you'll need to take a few dollars along to exchange for five or six rolls of pennies.

Ideally, you'll have a group of 10 pennies all minted before 1970, and one penny from each year between 1970 and 2000.

*You'll use an electronic balance scale to measure the weight of pennies from varying years.*

It's probably not a bad idea to get an extra roll or two, because you want to assure that you'll have enough pennies from both the control group and the variable groups to be able to conduct your experiment.

# Conducting Your Experiment

You've identified the problem you're attempting to solve, come up with a hypothesis, and gathered the materials you need; now you're ready to begin your experiment.

The experiment, as you know, is the heart of a science fair project, so be sure to work carefully and in an organized manner. Just follow these steps, and remember to carefully note your observations. It would be a good idea to fill in the charts illustrated in the next section, "Keeping Track of Your Experiment," as you proceed. You'll need to make a chart for each group of pennies to be able to record all your data. You'll end up with four charts.

Follow these steps:

1. Look through the pennies you have and find 10 that were minted before 1970. Do not include those made in that year.

2. List the date of each of the 10 pennies in chronological order on a data chart similar to the first one in the next section. Remember that the dates on this chart are just samples. You'll need to fill in the dates from the pennies you're using.

3. Using your balance scale, determine the mass of each penny in grams to the hundredths place, starting with the oldest and working up to the newest.

4. Calculate the average mass of one penny and record the mass on your data chart.

5. Select 10 pennies from your pile that were minted between 1970 and 1979.

6. Repeat steps 2 through 4, using those pennies.

> **Scientific Surprise**
>
> The average mass of one penny equals the total mass of 10 pennies, divided by 10.

7. Gather 10 pennies minted between 1980 and 1989.

8. Repeat steps 2 through 4, using those pennies.

9. Collect from your pile 10 pennies made between 1990 and 1999.

10. Repeat steps 2 through 4, using those pennies.

Once you've determined the average mass of each group of pennies, you'll be ready to begin analyzing your data.

# Keeping Track of Your Experiment

To keep track of your findings, use these charts or similar ones you make yourself. Remember that you'll have four charts when you're finished.

| Date of Penny | Mass in Grams |
| --- | --- |
| 1. 1955 | _____ |
| 2. 1956 | _____ |
| 3. 1960 | _____ |
| 4. 1961 | _____ |
| 5. 1962 | _____ |
| 6. 1964 | _____ |
| 7. 1965 | _____ |

| Date of Penny | Mass in Grams |
|---|---|
| 8. 1966 | _____ |
| 9. 1967 | _____ |
| 10. 1969 | _____ |
| average mass in grams: | _____ |

| Date of Penny | Mass in Grams |
|---|---|
| 1. 1970 | _____ |
| 2. 1971 | _____ |
| 3. 1972 | _____ |
| 4. 1973 | _____ |
| 5. 1974 | _____ |
| 6. 1975 | _____ |
| 7. 1976 | _____ |
| 8. 1977 | _____ |
| 9. 1978 | _____ |
| 10. 1979 | _____ |
| average mass in grams: | _____ |

| Date of Penny | Mass in Grams |
|---|---|
| 1. 1980 | _____ |
| 2. 1981 | _____ |
| 3. 1982 | _____ |
| 4. 1983 | _____ |
| 5. 1984 | _____ |
| 6. 1985 | _____ |
| 7. 1986 | _____ |
| 8. 1987 | _____ |
| 9. 1988 | _____ |
| 10. 1989 | _____ |
| average mass in grams: | _____ |

| Date of Penny | Mass in Grams |
|---|---|
| 1. 1990 | _____ |
| 2. 1991 | _____ |
| 3. 1992 | _____ |
| 4. 1993 | _____ |
| 5. 1994 | _____ |
| 6. 1995 | _____ |
| 7. 1996 | _____ |
| 8. 1997 | _____ |
| 9. 1998 | _____ |
| 10. 1999 | _____ |
| average mass in grams: | _____ |

When you've finished your experiment and have each of the four data charts filled out, you'll need to look at each chart and begin making some comparisons.

# Putting It All Together

The bottom line, of course, is whether the pennies minted pre-1970 are heavier than those in any other group. If they are, what conclusions can you draw? Remember those facts presented a few pages back in the section about reaching a hypothesis? Think about those facts—they'll help you to draw some interesting conclusions.

Also note any other interesting observations. For example:

◆ Did the weight of the pennies decrease steadily from one decade to the next?

◆ Is there one decade, or perhaps even one year, in which the mass changed significantly?

◆ Are the newest pennies the lightest ones?

◆ What's the difference of the average mass in grams between the heaviest group and the lightest group?

◆ Are there any kinds of patterns or disruptions to patterns?

**Standard Procedure**

The cost of copper increased dramatically during the 1980s, forcing the government to change the proportion of copper to zinc found in pennies.

Make all the observations you can, and use them to help you formulate a conclusion.

You could represent the data on your charts on one line graph. On the horizontal line (called the X axis), you would write the year of each penny you weighed, beginning with the earliest year.

On the vertical line (called the Y axis), you would list the range of masses from lightest to heaviest.

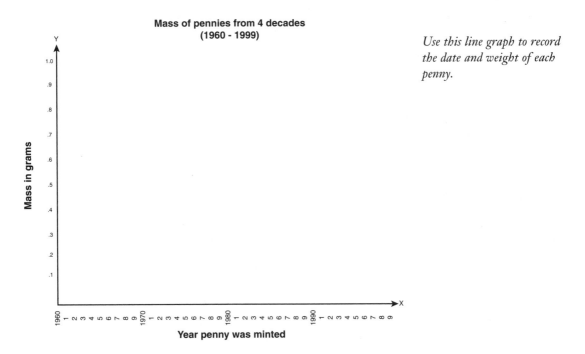

*Use this line graph to record the date and weight of each penny.*

Once you've graphed your information and studied your conclusions, you can come up with a decisive statement concerning the chemical composition of the penny. Are pennies made after 1970 and throughout the following decades lighter than those made prior to that year? Was your hypothesis correct?

# Further Investigation

As suggested earlier in this chapter, if you've determined that the weight of pennies has, indeed, been changing over the decades, maybe you'll want to take a closer look and try to figure out why.

The obvious answer seems to be that pennies are lighter than they used to be because they contain less copper and more zinc in an effort to offset the rising cost of copper. But, according to figures from AME Mineral Economics, a global firm of independent economists in the metal and mineral industries, the cost of copper actually declined in 2001. Does this mean the government might start replacing the zinc found in pennies with copper?

Another way to go a step further on this project is to repeat the experiment using 10 pennies from each year of the particular decade in which you noticed a significant change in the average mass of the pennies, compared to the control group. You would, for instance, determine the mass of 10 pennies dated every year between 1980 and 1990, meaning that you'll need 100 pennies.

> **Standard Procedure**
>
> You can check out what AME Mineral Economics has to say about the cost of copper and other minerals by going to its website at www.ame.com. You'll get a menu from which you can select the mineral you're interested in.

By doing so, you'd be able to tell if the mass decreased steadily each year, or if it was steady for several years and then took a big drop. You may notice some interesting patterns and be able to pinpoint a particular year in which the chemical composition changed significantly.

# Other Great Chemistry Projects

If you enjoyed learning about chemical composition and how the density of a material affects the mass of an object, you might have a special interest in chemistry and chemical compounds.

Two other chemistry-based science fair projects are described in the rest of this chapter. The information presented will give you a general idea of how to do the project, but doesn't walk you through all the steps as we did with the penny project. Don't forget that you'll need to follow the steps of the scientific method if you're going to work on one of these projects.

## Can You Be a Human Battery?

While the title of this science fair project sounds a little strange, it's actually pretty neat, and not difficult to do.

You'll need a piece of special equipment called a multimeter or a DC (direct current) microammeter. These measure electrical current in a circuit, and can be found at your local Radio Shack store for about $12.

In this experiment, you'll be your very own control. Your variables could be friends, or wet hands, or gloves, and so forth.

What you're trying to make happen is to have electrons travel through your body from one metal to the other. If you can do this, you're a human battery.

To do this, mount a piece of copper metal to a piece of wood, and a piece of aluminum metal to a different piece of wood. You can find these materials at a building supply store if they're not already in your garage.

Connect one end of the multimeter to the copper, and the other to the aluminum.

Your role in this experiment is to complete the electrical circuit from one metal to the other. By placing one hand on the copper and the other on the aluminum, the slightly acidic sweat on your hands provides the correct medium for this reaction to occur. You should see an electric current register on the multimeter. If you don't, reverse the connections and try it again.

Experiment a little bit now to find out how the current changes if you wet your hands before placing them on the metal. Doing so decreases the resistance to the flow of the electricity, causing the reading on the multimeter to be higher. You also can find out whether your friends or family members are better electrical conductors than you are.

> **Scientific Surprise**
>
> The chemical reaction that allows electrons to flow from the copper to the aluminum can't occur without an acidic solution, such as that found in a battery. In this experiment, you're providing the solution, therefore serving as the battery.

## What Materials Make the Best Crystals?

Growing crystals isn't difficult, and it makes a great science fair project. You can grow crystals on a sheet of black construction paper on a sunny day. You also can make them on a paper clip that's tied to a piece of yarn and suspended in a supersaturated liquid solution.

If you're going to use the construction paper method, you'll need to cut a piece of black paper so it fits into the bottom of a glass or metal pie plate. For the paper clip method, you can tie a paper clip to a piece of yarn, and then pull the paper clip and part of the thread through a piece of cardboard in which a small hole has been cut.

The key to either method of growing crystals is to make a supersaturated solution of water and salt or sugar. In this experiment, water is called the solvent, and the salt or sugar is called the solute. You can use regular old sugar, and either rock salt or Epsom salt. You probably have at least one of those materials available in your home. If not, you can buy any of them in the grocery store. Designate one solute as your control and the other two as variables.

> **Standard Procedure**
>
> We talk about "growing" crystals in this section, but you're not growing them in the strict sense of the word. You're causing them to form by supersaturating a solution of water and either salt or sugar.

Don't forget to state your problem, formulate a hypothesis, and follow the other steps of the scientific method.

To make a supersaturated solution, you simply need to add salt or sugar to hot water until the water is incapable of dissolving any more of the solute. That happens when the molecules of the solvent get so crowded together that there's no more room between them for solute molecules.

Start with a large glass jar or metal pot filled with two liters of very hot water. Working with one tablespoon (15 ml) at a time, add either the sugar or the salt to the water, stirring the solution well after each addition. Eventually, you'll see undissolved solute starting to collect at the bottom of the container. When this happens, your solution is supersaturated.

If you're using the construction paper method, pour enough of the supersaturated solution into the pie pan so that the paper is just covered. Set the pan in the sun and allow the water to evaporate. When it does, you should be able to see crystals formed in the bottom of the pan.

If you're using the paper clip method, fill a clean jar with the supersaturated solution. Then, place the cardboard over the mouth of the jar and secure the yarn to the cardboard so that the paper clip hangs in the middle of the jar. The bottom of the paper clip should be about five centimeters from the bottom of the jar.

As the hot solution cools and the water evaporates, the solute molecules will begin to cling to the yarn and crystallize.

You'll need to repeat the experiment three times, using all three solutes in order to see which one makes the best crystals. Photographing your results would be a good idea. And be sure to record your observations.

## The Least You Need to Know

- The problem solved in the course of this project is whether the chemical composition of pennies has changed during the past four decades.

- You need only a balance scale or other measuring device and pennies for this experiment.

- Keeping careful notes as you go along makes it easier to measure results and draw conclusions.

- Plotting your results on a line graph allows viewers and judges to quickly see and understand your results.

- Chemistry is a fascinating area of science that lends itself to fun and interesting science fair projects.

# Which Materials Insulate Best Against Windchill (and Other Great Earth Science Projects)?

## In This Chapter

◆ Understanding windchill and its effects

◆ Applying the concept of convection

◆ Encountering windchill in daily living

◆ Knowing what materials ward off windchill

◆ Working quickly to ensure accuracy

◆ Other fun earth science experiments

People love watching weather reports (The Weather Channel, anyone?) and talking about the weather. Storms sometimes are the top headline on the nightly news, with reporters outside covering them. The concept of windchill adds an extra dimension to weather-related conversations, and gives us added reason to complain when it gets very cold and uncomfortable outside.

In this chapter, you'll learn a little about windchill, and the materials that best guard against it. If you've ever been outside on a very cold, windy day, you know that this is information you'll be able to put to good use.

# So What Seems to Be the Problem?

We've probably all experienced windchill firsthand. Windchill is when your thermometer is telling you one thing, but your face and hands are telling you a different story.

| Scientific Surprise |
| --- |
| If the air temperature is above freezing, but the windchill is below freezing, a glass of water will not freeze. |

Stated a little more scientifically, windchill is a term that meteorologists use to describe how the air temperature feels, due to the effects of the wind. The windchill factor, according to our trusty dictionary, is the temperature of windless air that would have the same effect on exposed human skin as a given combination of wind and air temperature.

Below is a chart that shows real temperatures and "real feel" temperatures, which are how cold it feels with the windchill factored in.

**Windchill Chart**

| | If the temperature on the thermometer reads: | | | | | | | |
| --- | --- | --- | --- | --- | --- | --- | --- | --- |
| | 32°F 0°C | 23°F −5°C | 14°F −10°C | 5°F −15°C | −4°F −20°C | −13°F −25°C | −22°F −30°C | −31°F −35°C |
| **Wind Speed** | and the wind is blowing, then the "real feel" temperature would be: | | | | | | | |
| 6mph 10kmp | 28°F −2°C | 19°F −7°C | 10°F −12°C | 1°F −17°C | −8°F −22°C | −17°F −27°C | −26°F −32°C | −36°F −38°C |
| 12mph 20kmp | 19°F −7°C | 10°F −13°C | −2°F −19°C | −13°F −25°C | −24°F −31°C | −35°F −37°C | −45°F −43°C | −58°F −50°C |
| 18mph 30kmp | 13°F −11°C | 10°F −17°C | −11°F −24°C | −24°F −31°C | −35°F −37°C | −47°F −44°C | −58°F −50°C | −71°F −57°C |
| 24mph 40kmp | 9°F −13°C | −4°F −20°C | −17°F −27°C | −29°F −34°C | −42°F −41°C | −54°F −48°C | −67°F −55°C | −80°F −62°C |
| 30mph 50kmp | 5°F −15°C | −8°F −22°C | −20°F −29°C | −33°F −36°C | −47°F −44°C | −60°F −51°C | −72°F −58°C | −87°F −66°C |

*Windchill chart.*

We hear much more about windchill than we used to, but it's by no means a new concept. The term "windchill" was coined in 1939 in a dissertation called "Adaptation of the Explorer to the Climate of Antarctica."

Written by Paul Siple, an Antarctic explorer, the dissertation explored the relationship between wind and outdoor comfort level, among other things. In the years following the dissertation, Siple and fellow explorer Charles Passel experimented with how long it took to freeze water that was exposed to the elements.

| **Scientific Surprise** |
|---|
| Siple's original work on windchill was used until 2001, when the U.S. National Weather Service and the Canadian Weather Service replaced his formula with an updated one. The new formula was based on increased knowledge and experimentation. |

You probably can guess what they discovered—that the time needed to freeze water depended on the temperature of the water to start, the outside temperature, and the wind speed. That work spawned the windchill factor, which is still calculated and used today.

The windchill factor is not an actual temperature, but a measure of how quickly heat is lost from an object. An algebraic formula is used to determine windchill. The windchill formula was revised and implemented by the National Weather Service in 2001. To see the formula and learn more about windchill, check out this website: www.usatoday.com/weather/resources/basics/windchill/wind-chill-formulas.htm.

The problem you'll be attempting to solve by doing the experiment in this chapter is which materials insulate best against windchill. Using common materials probably already in your house, you'll check out which ones protect water exposed to wind. In this case, the wind will come from a fan. That way, you can do the experiment any time you want, and you don't have to hang around outside on a cold, windy day.

If you want to, you could use the name of the chapter, "Which Materials Insulate Best Against Windchill?" as the title for your science fair project. Other titles to consider include:

♦ Protecting Against the Big (Wind) Chill

♦ Warding Off Windchill

♦ How Can You Avoid the Effects of Windchill?

In the next section, we'll take a look at why it's important to have an understanding of windchill and which materials can protect against it.

# What's the Point?

While some scientific concepts might seem a little abstract and not especially applicable to everyday life, the idea of windchill is something nearly everyone can relate to. Windchill not only affects how we need to dress and our outdoor comfort level, it applies to other daily activities, as well.

For instance, how many times have you sat down to eat a bowl of soup, or drink a cup of hot chocolate or hot tea, only to find out that it's too hot to put in your mouth?

When that happens, what do you do? If you're like a lot of folks, you blow on the soup or the beverage to make it cool enough to eat or drink. Let's think about why blowing on your soup makes it cool faster than simply letting it sit on the kitchen table for a few minutes.

When you blow on your soup (and, please, don't ever blow on anyone else's soup—it's considered rude!), you're simulating the action that the wind performs on your face and hands on a cold, windy day. By blowing on the soup, you increase the movement of the air around it, which lets heat escape from the surface of the soup more quickly.

Not only does this air cause the soup to cool more quickly, it causes an interesting action to occur within the bowl (who would have thought that a bowl of soup could present so many scientific opportunities?). As heat escapes from the top layer of the soup, the slightly cooler, and now slightly denser, liquid sinks to the bottom of the bowl. This causes the hotter, and less dense liquid to rise to the top of the bowl, creating a convection current. This process is called *convection*.

> **Basic Elements**
>
> **Convection** is the transfer of heat by the movement of currents within the heated material.

It's the windchill, caused by you blowing on the soup, that causes the soup to cool—just as it's windchill on your skin that makes it feel colder outside than indicated on the thermometer.

Often, instead of worrying about making soup or coffee or hot chocolate cooler, people worry about keeping it hot. Restaurants that pack food and drinks to go are careful to enclose it in material—usually Styrofoam—that will help to keep the food hot until you eat it. If you ever got a cup of hot chocolate or some soup from a take-out restaurant, chances are it was in a Styrofoam cup or bowl.

In your experiment for this science fair project, you'll test five containers made of different materials to determine which would insulate best against windchill and keep your soup or hot chocolate the hottest. You'll find out whether Styrofoam, as most restaurants seem to think, is the best insulator.

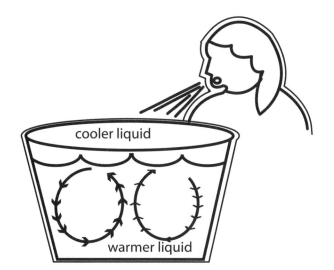

*Blowing on a bowl of hot soup transfers heat within the bowl by creating moving currents.*

It's important to get a good assortment of cups. Those recommended to use are an insulated coffee mug; a regular, ceramic coffee mug; a glass cup; and a plastic cup. You'll test which container protects best against windchill by pouring boiling water into each cup, inserting a thermometer, and setting all the cups in front of a fan.

You'll know which cup insulates best against the windchill caused by the fan because that water will stay the hottest for the longest time.

The Styrofoam cup will be the control, and the other types of cups the variables.

By seeing which type of cup insulates best against windchill, you'll learn whether those restaurant folks really know what they're doing.

**CAUTION**

### Explosion Ahead

Because you're working with boiling water, you'll need to be extremely careful when conducting this experiment. Boiling water will cause burns if spilled on your skin. Be sure to ask an adult if you need help.

# What Do You Think Will Happen?

You probably already have a good idea of which types of containers keep liquids hot and which don't. So you should be able to make an educated guess—or hypothesis—about how quickly the water in each cup will reach room temperature.

If you trust that restaurants and packaging companies know what they're doing, you might guess that the Styrofoam cup will protect best against the windchill caused by the blowing fan. Or you might think one of the other cups will do a better job.

You will come up with a quantitative measurement for each cup tested. Quantitative means to describe a property of an object using numeric measurements. Examples would be the amount of water used, the temperature of the water and the time it takes for the water to cool down to room temperature. The words amount, temperature and time all imply a number value with these terms. These properties are all quantitative.

You can record your results on the data chart in the section of this chapter called "Putting It All Together."

# Materials You'll Need for This Project

You can use practically any type of fan for this experiment, and a window-type air conditioner that blows cold air inside would work, as well. You just need to make sure that you can place the cups at a level where they'll receive the full effects of the fan when it is turned on.

You will need five thermometers for this experiment, preferably the same type. You can find these at a hardware or home-improvement type store. Don't use fish aquarium thermometers. They don't measure to 212 degrees Fahrenheit (100 degrees Celsius), which is the temperature at which water boils.

**Standard Procedure**

The thermometers you'll use will have to have a range from about 50 degrees Fahrenheit (10 degrees Celsius) to 212 degrees Fahrenheit (100 degrees Celsius).

You'll need a tray to hold the five cups; make sure it's large enough and sturdy enough.

You can boil the water you'll use in a pot or teakettle on the stove, or in the microwave. If you use the microwave, make sure you can heat at least 24 fluid ounces (720 ml) all at once. You'll be pouring six ounces (180 ml) of the boiling water into each of the five cups.

The materials you'll need for this experiment are listed below:

- A fan
- Eight-ounce (240 ml) Styrofoam drinking cup
- Eight-ounce (240 ml) glass drinking cup
- Eight-ounce (240 ml) plastic drinking cup
- Eight-ounce (240 ml) ceramic mug
- Eight-ounce (240 ml) metal insulated cup
- A tray large enough to hold the cups

- Five thermometers
- A one-cup glass measuring cup
- Twenty-four fluid ounces (720 ml) of boiling water
- A clock or watch with a second hand

When you've collected all the materials, you'll be ready to begin your experiment.

# Conducting Your Experiment

Follow these steps:

1. Place the fan on the floor or on a table.

2. Check and record the temperature in the room so you'll know when the water in the cups has reached room temperature.

3. Place the five cups side-by-side on the tray.

4. Place the tray 18 inches (45.7 cm) away from the fan. Note: All of the cups must be on the same level as the fan, so that the wind blows directly on them.

5. Pour six fluid ounces (180 ml) of boiling water into each cup.

6. Immediately place a thermometer in each cup and record the temperature of the water. Record these temperatures in a data chart similar to the one in the next section, "Keeping Track of Your Experiment."

**Standard Procedure**

It's a good idea to have somebody help you pour the water into the cups so that the procedure is finished quickly. You don't want the first water you poured to get a head start on cooling.

*Use five different thermometers to measure the temperature of the water in each cup.*

7. Turn the fan on to its highest setting, and write down the time that you turned it on. This is the starting time for your experiment.

8. Check the thermometers frequently, watching carefully at what time the water in each cup reaches room temperature. Record that time for each cup.

9. Repeat steps 1 through 8 two more times. Doing three trials will give you much more accurate results than just doing the experiment once. You'll have to make two more data charts on which to record the results of the additional trials.

Work carefully and remember to keep checking the temperature of the water in each cup.

# Keeping Track of Your Experiment

If you do this experiment three times, you're going to have a lot of numbers to record. Use a chart such as the one below to record the results of each of the three trials.

**Room temperature** ———— **degrees Fahrenheit**

———— **degrees Celsius**

| Type of Container | Styrofoam Cup | | Glass Cup | | Plastic Cup | | Ceramic Cup | | Metal Insulated Cup or Mug | |
|---|---|---|---|---|---|---|---|---|---|---|
| Initial temperature of hot water | °F | °C | °F | °C | °F | °C | °F | °C | °F | °C |
| Final time experiment started in minutes | | | | | | | | | | |
| Final time when water reached room temperature | | | | | | | | | | |
| Total time needed for water to cool to room temperature | | | | | | | | | | |

*Recording your results.*

# Putting It All Together

Once you've completed your experiment and recorded your data, it's time to definitely and decisively proclaim the best cup for protecting against windchill. You'll be giving a quantitative analysis, based on the times recorded on your data charts.

If you did three trials, you'll need to average the times you got for each cup. You do this by adding the three times together, and then dividing by three. That number is your average time.

So which of the five cups insulated best against the effects of the windchill produced by the fan? How does your hypothesis measure up? Were you right?

# Further Investigation

If you enjoyed this project and want to follow up a little bit, a couple of ways you can extend it or vary it are listed below.

◆ Compare the effects of windchill on several types of insulated cups of different brands. This would allow you to compare the same kind of cup, but find out which brand works best.

◆ Use six of the same kind of cup (say, Styrofoam), but wrap five of them in different materials such as felt, silk, aluminum foil, wool, and so forth. Keep one container unwrapped to serve as the control. Follow the same procedures as described in the main experiment, and you'll discover which materials most help to retain the heat of the water.

# Other Great Earth Science Projects

If you enjoyed this science fair project and would like to get an idea for another that is earth science–oriented, read the ideas presented below. One deals with clouds and weather, and the other explores which factors most affect the rate of evaporation.

Earth science is an interesting area of science because it is very much a part of our everyday lives. Maybe you admire the mountains in the distance, and are prompted to study the different types of rocks that make up those fascinating formations. That's earth science.

Or perhaps you love to look at the night sky, using a telescope to locate other planets and star formation. That, too, is earth science.

Earth science is the study of the earth, which includes the solid earth, oceans and lakes, the atmosphere and the universe. Questioning the processes that make our planet tick can be a daily, nearly constant, activity.

## Do Clouds Tell Us What Weather Is Coming?

Do you know anyone who seems to be able to check out the sky and predict what the weather will be? If so, how do you think he or she does so?

Observing clouds can tell us a lot about what kind of weather to expect. The experiment described in this section will help you to know what to look for when using clouds to predict upcoming weather.

**Standard Procedure**

An experiment such as this would require at least several weeks to complete. If you're interested in doing something similar, be sure that you start well in advance of your project's due date.

**Standard Procedure**

It would be a good idea to take photographs of various types of clouds and include them, clearly marked, in your display.

There are several types of clouds that appear due to approaching and current weather conditions. You probably know what thunderstorm clouds look like, and the kind of weather they bring.

Cirrus clouds are high and feathery, and indicate good weather. Other types of clouds, such as stratus and cumulus, can be indicators of rain or snow. If you're interested in learning how to predict weather by watching the clouds, you'll need to do some research first. Check out some websites or grab a book from your school library for more information.

To begin the experiment, record your starting date, time, and the temperature on a chart in a journal. Begin observing the clouds twice a day, writing down your observations each time. Try to observe them at about the same times each day, and note specific information such as cloud color, type, approximate wind speed and direction, and so forth.

If you check out the sky at night, record anything interesting such as a ring round the moon. Write down as much about the weather as you can, along with your observations of the sky.

Once you have collected several weeks' worth of data and observations about the clouds you've seen, objectively interpret your results. Were there any significant patterns related to cloud type and weather that developed?

With the knowledge you've gained about clouds and weather, do you think you could predict the day's weather without seeing it first on television or reading about it in the newspaper?

## What Factors Most Affect Evaporation?

Evaporation of a substance such as water occurs when the molecules in a liquid state absorb enough heat energy to vaporize into a gas. The heat energy, which causes the molecules to move, is called kinetic energy.

The water molecules remain the same during this change, except that they're moving much more quickly and their energy increases due to that movement.

If you remove the energy by cooling the gas, then the gas condenses back into a liquid. Take away even more of the heat energy, and the liquid will solidify. This is what causes water to turn to ice.

You've probably noticed that puddles you see in front of your house in the morning often are gone by the time you get back from school. What happens to them? The heat from the sun causes the water to evaporate—or turn to gas. It takes longer for water to evaporate in cool weather than hot weather, and the deeper the puddle, the longer it will take for all the water to disappear. Wind can also be a factor affecting evaporation.

If this sounds interesting to you, you can devise an experiment to test the different factors that affect the evaporation rate of water.

You'd simply put a little water into two shallow containers, such as pie pans. Don't use a lot of water, or it will take too long for it to evaporate. About 2 teaspoons (10 ml) will do. The water in both containers should be the same temperature.

Create variables, such as putting one pan directly in the sun, and the other in the shade. Or put one pan in front of a running fan, and the other in a still place. You'll need a watch or clock with a minute hand so you can time how long it takes the water in each circumstance to evaporate.

Try putting some water on the top of a plate to see if the surface area of the water makes a difference in the evaporation rate.

You'll gather a lot of quantitative information in this science project. Keeping all your information together in a journal would be helpful. Conducting three trials for every variable tested is always a good idea in order to achieve more accurate results.

## The Least You Need to Know

◆ Windchill is a term used to describe how the air temperature feels due to the effects of the wind.

◆ Nearly everyone can relate to windchill because we encounter it often in everyday life.

◆ Knowing which materials best insulate against windchill has practical, usable applications.

◆ It's important to work quickly in this experiment in order to keep all the water at the same starting temperature.

◆ Some other earth science projects involve using clouds to predict the weather, and the evaporation rate of water.

# Chapter 16

# What Materials Conduct Static Electricity Best (and Other Great Physical Science Projects)?

## In This Chapter

- ◆ Understanding static electricity
- ◆ The world of protons, neutrons, and electrons
- ◆ Transferring electrons to create a charge
- ◆ Building an electroscope
- ◆ Comparing how different materials take or give electrons
- ◆ Other physical science experiments

There's little doubt that you've had some experience with static electricity. That annoying "shock" you feel after you walk across the carpet and touch the doorknob, for instance. Or that equally annoying thing your hair does when you pull a wool hat quickly off your head. But what is static electricity, and why does it cause these types of events to occur?

In this chapter, you'll not only learn about static electricity, but also about a really cool experiment regarding static electricity that you can do with a simple device you'll make called an electroscope. You'll use the electroscope to detect electrical charge.

## Scientific Surprise

There are only 92 naturally occurring elements, and 16 others that are synthetic, or man-made. Each of these elements has different numbers of protons, neutrons, and electrons. Everything around us is made up of combinations of these elements.

## Basic Elements

**Protons** and **neutrons** are tiny particles contained within the nucleus of an atom. **Electrons** are even tinier particles that orbit around the nucleus.

# So What Seems to Be the Problem?

The problem you'll attempt to solve is how well certain objects give up their electrons to other objects. How does that relate to static electricity, you ask?

In the wonderful world of science, there are *protons*, *neutrons*, and *electrons*. Protons and neutrons are tiny particles contained within the nucleus of an atom.

An atom, if you'll recall, is the smallest piece possible of an object. A bar of silver, for instance, could be divided in half, then half again, and again and again and again, until there is a piece so small that if it were to be divided, it would no longer be silver. That very last piece is called an atom.

In the middle of each atom is a nucleus. That's where the protons and neutrons hang out. Electrons, on the other hand, are even smaller than protons and neutrons, and orbit around the nucleus of the atom.

## Electrical Charges

Protons, neutrons, and electrons are very different from one another. In terms of this experiment, the way that they're different from one another concerns their electrical charges.

Protons have a positive (+) charge. Electrons have a—you guessed it—negative (–) charge, and neutrons, as their name suggests, have no charge. When an atom contains the same number of protons and electrons, the atom has no overall charge, but is neutral. That's because the positive charge of a proton is equal to the negative charge of an electron. Put together, one cancels out the other.

While protons and neutrons stick closely together within the nucleus of an atom, electrons are like free spirits who can't stay still. They move around about the nucleus, and sometimes bail out on the atom altogether and move to a different atom.

When this happens, it puts the electric charge of the atom out of balance. Remember, a neutral atom must contain the same number of protons and electrons. When electrons jump ship and move to another atom, the balance is lost.

An atom that loses electrons then has a positive charge because it contains more protons than electrons. The atom that gains the electrons has more negative than positive particles, so it has a negative charge.

Just to make things a bit more confusing, an atom that has either a positive or negative charge is no longer called an atom. It's now known as an ion.

Some materials hold their electrons very closely, not allowing the electrons to move through them well. These materials are known as insulators. Materials that do allow their electrons to move through them easily are called conductors. Cloth is a good insulator, while metals generally are good conductors.

## Electrons and Static Electricity

It's important to realize that electrons moving from place to place is not an unusual occurrence. It happens all the time, whenever two objects rub together. When there is a lot of contact between two objects, a lot of electrons get transferred, and the amount of charge builds up. *Static electricity*, simply put, is nothing more than an imbalance of positive and negative charges.

The next idea to understand is that opposite charges attract, while charges that are the same repel each other. When you pull a hat off your head and your hair does that weird, standing-up-straight thing, it's because of this attract-and-repel rule.

Electrons that were in your hair rub off onto the hat. When you remove the hat, the electrons go with it, leaving your hair with only a positive charge. Each little hair tries to get away from its same-charged neighbor, resulting in that flyaway look.

**Basic Elements**

Static electricity is an imbalance of positive and negative charges.

All right, enough explanation. It's time to get down to business. In this science fair project, you'll attempt to discover which objects most readily release their electrons, allowing a charge to result. Remember that some objects (insulators) do not give up electrons as easily as others (conductors).

If you want to, you can use the title of this chapter, "What Materials Conduct Static Electricity Best?" as your project title. Or you could call your project one of the titles suggested below:

♦ How Does Transferring Electrons Change a Charge?

♦ Understanding Static Electricity

♦ Using an Electroscope to Detect Electrical Charge

Once you understand the concept of static electricity, you can think about everyday applications of this project.

# What's the Point?

Now you know why your hair stands straight up when you pull a wool hat off your head. So what?

Understanding electrical charges is important for many reasons. It helps you to understand how the world works, and why certain events occur.

Did you ever notice that static electricity shocks occur mostly in the winter? And that even if you don't pull a hat off your head, your hair tends to be a little flyaway when it's cold outside?

This is because the air in the winter generally is very dry. Summer air normally is more humid, meaning that it contains more moisture. What happens is that the water in the summer air helps electrons to move off your body more quickly. Water is a good conductor.

Because the electrons move off your body, you don't build up a big charge. In the winter, however, when the electrons stay on your body, you build up a negative charge. So when you walk across the rug, the electrons from the rug move onto, and build up on, your body. When you touch the doorknob, the electrons jump from you to the metal knob (remember that metal is a good conductor), causing you to feel a shock.

In this experiment, you'll use the electroscope you'll make to determine which objects best build up and conduct electrical charges. The control you'll use is uncharged aluminum foil, and the variables you'll use will be other common objects, all of which are normally found around a house.

| Scientific Surprise |
| --- |
| Charged objects emit invisible electric force fields that surround them. The strength of these fields varies depending on many factors. Way back in the 1780s, a scientist named Charles Coulomb described and investigated these force fields. The formula he came up with for figuring out the strength of the fields is called Coulomb's Law. |

# What Do You Think Will Happen?

You should have a basic but sound understanding now of how static electricity occurs, and what happens when it does. Before you venture a hypothesis, though, you'll need to understand how the experiment works.

In this experiment, you'll test how well certain objects transfer electrons, using aluminum foil as the detector.

During the experiment, you'll transfer electrons from one object to another by rubbing a flexible, plastic ruler on different materials. The ruler will serve as the conductor of the electrons onto the aluminum foil.

When you rub the ruler on different objects, it will either pick up electrons from the object, or it will pass some of its electrons onto the object. If the ruler picks up electrons from the material on which it's rubbed, it will have a negative charge. If the ruler gives off electrons, it will have a positive charge.

Your job is to see how the uncharged aluminum foil (the detector) reacts to the ruler (the conductor) after the ruler has either gained or lost electrons. The materials you'll test are wool (a wool sweater will work well), a piece of silk (such as a scarf or a tie), cotton fabric (a pillowcase, perhaps), newspaper, and carpet (you know where that is).

To come up with a hypothesis, venture a guess on which materials you think are most likely to transfer electrons onto the ruler. Do you have a hunch that some might be more effective at doing that than others?

Go ahead and take your best guess. Then we'll get started on the experiment.

**Standard Procedure**

There are many fun experiments you can do with static electricity. For a bunch of ideas and information concerning static electricity, check out this website: www.eskimo.com/ ~billb/emotor/statelec.html.

# Materials You'll Need for This Project

An electroscope is a device that detects electrical charge. There are many types of electroscopes, some complicated models that you could buy from a scientific supplier, and some simple ones you can make yourself.

You'll need only four materials to make a simple electroscope. They are:

◆ One small cup (glass or paper)

◆ One plastic drinking straw with flexible end

- Tape

- Aluminum foil

The other materials you'll need for the experiment are those mentioned in the previous section:

**Standard Procedure** _____

This experiment will work best on a day that is cool and dry, rather than warm and humid. Conducting it inside the house on a day when the heat is turned on would give you favorable conditions.

- A flexible plastic ruler

- Wool

- Silk

- Cotton

- Newspaper

- Carpeting

If you can't get the materials specified above, feel free to substitute something you have on hand. Part of the beauty of this experiment is that you probably won't have to buy anything. If you do, it shouldn't cost more than a dollar or two to get what you need.

# Conducting Your Experiment

Before you begin the actual experiment, you'll need to build your electroscope. Don't worry—"building" the electroscope really doesn't require any building.

*Electroscope.*

Follow these steps:

1. Place a straw into a small cup. The flexible part of the straw should be at the top, and the straw should be bent.

2. Cut two small strips of aluminum foil, about 2½ inches (6 cm) long, and ½ inch (1 cm) wide.

3. Tape the strips onto the bent part of the straw, so that they're next to each other, but not touching. They should hang straight down from the bent arm of the straw.

   Now your electroscope is ready and you can move on to the next steps of the experiment.

4. Rub the ruler onto the piece of wool, and then bring the ruler close to the aluminum foil, without actually touching it.

5. Record your observations, noting which material you used and the reaction you observed with the foil strips.

6. Repeat steps 4 and 5 with each of the other materials. Note which materials result in the aluminum foil being attracted to the ruler, and which make the foil and ruler repel one another. Realize that the neutral foil strips may at first be attracted to the charged ruler, and then within a second or two repel the ruler. This would occur because the foil strips picked up electrons from the ruler, and both have the same charge.

7. Repeat the entire experiment three times, but leave an hour in between each repetition to let the foil strips regain some stability. Try to make sure the surroundings for the experiment are the same for each trial.

> ### Scientific Surprise
>
> Rubbing the ruler with silk will remove electrons from the ruler, giving the ruler a positive charge. Wool, however, will transfer electrons onto the ruler, resulting in a negative charge. You can test any material to see if it takes or gives electrons.

Be sure to keep accurate notes on what you observe with each material.

# Keeping Track of Your Experiment

You can use the following chart to keep track of your observations. Or you could make your own, similar chart, if you prefer.

| Ruler Rubbed With | Did the Foil Strips Repel or Attract Each Other? | Charge on the Ruler | Charge on the Silk | Other Observations |
|---|---|---|---|---|
| silk | | | | |
| wool | | | | |
| cotton | | | | |
| newspaper | | | | |
| carpet | | | | |

*Electroscope observations.*

# Putting It All Together

Once you've tested a variety of materials to see which conduct static electricity the best, you'll have a good idea about which materials release electrons and which take electrons.

Check your observations carefully to see which materials reacted best with the aluminum foil.

# Further Investigation

If you enjoyed this project, there are many more experiments you can do to explore static electricity and the transfer of electrons.

A fun experiment would be to test the carpeting found in different rooms of your house. See how a ruler rubbed onto the carpet of one room reacts with the aluminum foil, compared with that in other rooms.

You could do an informal experiment by walking across different carpets and then touching a metal object, such as the doorknob. You might be surprised to see a big difference in the amount of electrons moving from the different carpets onto you as you move across the room.

Another fun experiment is to see if you can build up enough static electricity on a balloon or other object to "bend" water. Simply turn on the water so that it's coming out of the faucet in a very thin stream.

Rub a balloon onto a wool sweater, stuffed animal, or other furry object, and then hold it close to the stream of water. The water, which is neutral, will be attracted to the charged balloon, and will "bend" toward it.

> **Scientific Surprise**
>
> No electrons are made or destroyed when something is charged with static electricity. They're merely transferred from one place to another, with the total electric charge remaining constant. This is called the principle of conservation of charge.

# Other Great Physical Science Projects

Physical science is the study of the world around us. It encompasses many areas, and is ever changing and expanding as scientists learn more and more about how the world is made and how it works.

If you're interested in physical science, check out the suggested experiments below.

## Can Water Make a Bottle Rocket Fly Higher?

We think of rockets as being fairly modern inventions, and they are. Germany was the hub of early rocketry in the 1930s. The first rocket airplane was developed in the United States in 1947—the first plane ever to exceed the speed of sound in level flight. John Glenn orbited the earth in 1962 in the Atlas rocket.

Although rockets are modern, the science of rockets, based on Isaac's Newton's third law of motion, has been understood for hundreds of years.

This law pertains to rockets, which are forced upward by burning fuel that produces large amounts of gases expanded by heat. You can create a "rocket" from a soda bottle, and fuel it with air pressure. The problem you'll attempt to solve is whether additional rocket fuel—in this case water—causes your rocket to fly better than just air pressure.

> **Scientific Surprise**
>
> Isaac Newton's third law of motion claims that for every action in nature there is an equal and opposite reaction.

You'll need a clean, plastic two-liter bottle, a bicycle pump with an air pressure gauge, some water, a rubber stopper to fit firmly inside the neck of the bottle, an inflating needle like those used for inflating basketballs, and a cardboard carton with a hole to support the neck of the bottle when turned upside down.

You'll need to do this experiment outside, and not too close to your or your neighbor's house.

With an adult to help, drill a hole in the rubber stopper into which you can fit the inflating needle. Attach the needle to the pump. Fit the stopper and nozzle firmly into the neck of the bottle. Place the bottle upside down on the cardboard carton, which will serve as your launch pad.

Pump air into the bottle until the air pressure inside forces the stopper out of the bottom, causing it to fly into the air. It will be difficult to measure how high the bottle flies up, so your results may have to be more qualitative than quantitative. You could try clocking the amount of time the bottle stays in the air, giving you an indication of its time in flight. Keep track of the air pressure and approximate height by recording those numbers in a data chart.

Repeat the experiment, but this time add one cup of water (354 ml) to the bottle. Pump air into the bottle until the air pressure forces the stopper out of the bottom. This time, water and air are forced down through the neck. This will create a force that will push the bottle up. Record the air pressure and approximate height of the bottle.

Try different amounts of water to determine which amount makes the bottle rocket go the highest.

## Downhill Discoveries

The concepts of potential energy, kinetic energy, drag, friction and acceleration come into play in this suggested science fair project, making it a valuable learning experience.

By making your own sloping tracks and selecting course variables, you'll be able to determine how course conditions can affect acceleration and race results. Think bobsledding, luge racing, and downhill skiing.

All the necessary materials can probably be found around your house, with the exception, perhaps, of the large cardboard box. Different materials used on the strips of paper or foil represent different conditions racers face.

You'll need a large piece of cardboard, such as from an appliance box, tape, a small fan, salt, water, cornstarch, butter, a meter stick or tape measure, aluminum foil or waxed paper, clay, a stopwatch, and five checkers.

Simulate trails by taping six strips of foil or waxed paper onto the cardboard, leaving some space between them. Don't fasten the strip at the bottom. Leave one strip as is—that's your control. On the other strips, sprinkle or coat with one of the following: salt and water mixture, cornstarch, water, and butter. The leftover strip will have a fan blowing onto it.

Prop up the cardboard to simulate a hill, and have one person stand at the top of the cardboard with a checker. Slide the checker down one path, recording distance traveled and time needed.

Use one checker for each of the other strips, recording time and distance traveled for each condition. Repeat procedure a total of three trials, using the clay to add weight to the checkers as another variable.

## The Least You Need to Know

- ◆ Protons and neutrons are tiny particles contained within the nucleus of an atom, while electrons move around outside the nucleus.

- ◆ Electrons are easily transferred from one object to another, resulting in static electricity.

- ◆ Opposite charges attract one another, while like charges repel one another.

- ◆ An electroscope is a device used to detect electrical charge.

- ◆ Some materials release electrons onto a conductor such as a ruler, while others take electrons from the ruler.

- ◆ Physical science is a wide-open field that explores many avenues of science.

# Part 4

## Advanced-Level Science Projects

Science is exciting at all levels, but becomes particularly intriguing when you begin to explore topics at a slightly higher level.

Chapters 17 through 21 introduce you to projects that deal with timely, fascinating topics such as DNA, vegetative cloning, and the differences between various types of plastics.

The step-by-step experiments and explanations of the topics allow you to delve into a project with confidence, and have a lot of fun learning as you go. Take your time and work carefully through these projects, using the ideas presented in the "Further Investigation" sections to follow up on your results.

# Is the DNA of a Cow Different from a Chicken's (and Other Great Biology Projects)?

## In This Chapter

◆ Understanding the basics of DNA

◆ Extracting and comparing DNA

◆ Isolating DNA

◆ Making qualitative observations

◆ Finding DNA in other materials

◆ More biology projects

You probably know a little about DNA, especially if you've chosen to do this project for your science fair. DNA has been a hot topic over the past few years as it relates to DNA patenting, genetic engineering, genes, and biotechnology.

You'll read a little bit about DNA before we move on to the actual experiment for this project. It's important to have a good understanding of what

DNA is and how it works before thinking about the particular experiment outlined in this chapter.

So get ready to learn a little more about DNA, and then you'll learn how you can extract DNA from cow and chicken livers in order to compare them.

# So What Seems to Be the Problem?

The problem, as stated in the title of this chapter, is to find out whether the DNA of a cow is different from the DNA of a chicken. To solve that problem, you need to know something about DNA, and how it varies from species to species.

### Basic Elements

**DNA** is short for deoxyribonucleic acid. It is the material that determines inheritance of eye and hair color, height, stature, and so forth. DNA is found in every cell of most living organisms, including humans, animals, fish, birds, plants, and bacteria. It's even present in viruses.

### Scientific Surprise

Among different types of proteins are structural proteins that make up hair, fur, tails, horns, fingernails, spiderwebs, tendons, ligaments, and connective tissue; storage proteins such as seeds and eggs that store amino acids for developing plants and animals; enzymes, which speed up chemical reactions; and contractile proteins, which are the major component of muscles.

*DNA*—short for deoxyribonucleic acid—is a very special molecule that is necessary to life. All living organisms contain DNA. It is the material that determines what color eyes we'll have, what color our hair will be, how tall we'll be, and many other human, plant, and animal traits.

The bodies of animals, including humans, are made up of different types of cells. These cells include muscle cells, blood cells, skin cells, bone cells, nerve cells, and many others.

DNA contains a chemical code that's used to make proteins within these cells. DNA transmits chemical messages from the nucleus of a cell to other molecules, specifically RNA (ribonucleic acid). This, in turn, builds a polypeptide chain of amino acids that become a protein.

There are 20 different amino acids. The different combinations of these amino acids in a polypeptide chain determine the function of the protein.

There are both proteins and DNA found within the nuclei of every cell of an organism.

Proteins are large molecules that serve many different functions. There are different types of proteins, all of which perform different jobs within a cell.

During the course of the experiment in this chapter, you'll learn how you can release, or extract, DNA from its nucleus. Once you've done that, you can compare the DNA from the chicken liver with that from the cow liver to see if they're different.

If you want to, you can use the title of this chapter, "Is the DNA of a Cow Different from a Chicken's?" as the title of your science fair project. Or you could consider one of the titles suggested here:

◆ Extracting and Comparing DNA of Different Animals

◆ Unlocking the Secret of DNA

Whatever title you choose for this project, be sure to work carefully, and follow the directions closely. This project is a little more complicated than most in this book, and will require careful work.

# What's the Point?

DNA sounds like a terribly complicated, mysterious matter, but it's basic to every living thing. You are what you are, largely, due to DNA. Understanding DNA also allows scientists to study genetically related diseases and look for means of preventing them.

The point of this science fair project is to show you that DNA isn't mysterious or something that only scientists can comprehend. You'll actually isolate and retrieve DNA from two different animals in order to examine and compare it.

| Scientific Surprise |
| --- |
| James Watson and Francis Crick received the credit for determining the double helix shape of DNA in 1953 in Cambridge, England. |

This is a multi-step process, but it's not overly difficult or complicated. Knowing how it occurs will give you a much better understanding and appreciation for what happens in biology labs.

# What Do You Think Will Happen?

Before you venture forth with a hypothesis, you should have a basic understanding of what you'll be doing during the experiment.

Don't get grossed out, but you'll be working with chicken and cow livers, each of which contain many cells, and in turn, DNA and proteins. The first thing you'll do is chop up the liver in a blender. This will begin to break up the cells.

*You'll conduct an experiment using liver cells to determine whether the DNA of a cow is different from that of a chicken.*

Then, because you want to get at the DNA contained within those cells, you'll need to break the plasma or cell membranes that surround cells. This process is known as *lysing*. Plasma membranes are constructed of two layers of compounds, called phospholipids.

### Basic Elements

The process of breaking the plasma of the cell is known as **lysing**.

Phospholipids act like fats, in that they don't mix with water. This fatty substance is located on the inside, but not the outside, of the membrane.

The plasma membrane can be broken apart by using liquid dishwashing soap. This acts as a "degreaser," cutting through the fatty material in the membrane.

The dishwashing liquid also helps to break apart the nuclear envelope, which surrounds the nucleus where the DNA is located. Once you break apart the membrane and the nuclear envelope, the DNA will be exposed, allowing you access to it.

*An animal cell containing DNA.*

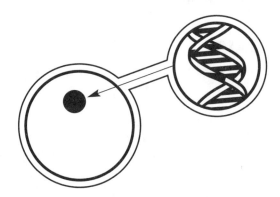

In order to remove any proteins that are still entangled in the DNA, you'll treat the DNA with an enzyme such as meat tenderizer or enzymatic contact lens cleaner. This ingredient will be sparingly added to the DNA mixture.

The next step is to isolate the DNA so that you can retrieve it and examine it. You can do this by mixing an equal amount of cold rubbing alcohol to the DNA mixture.

DNA is insoluble in rubbing alcohol, and it will separate out of the solution. The mixture will now have three distinct layers: a bottom liquid mixture of broken cell components, a middle layer of stringy DNA and a top layer of rubbing alcohol.

Once you've got the separated material, you can retrieve the DNA by spooling it onto a stick.

If you're thinking this sounds a little bizarre, just stay with us for a little bit longer. This isn't as difficult or complicated as you might think.

Once you process the DNA from both the cow and chicken livers, you'll be able to observe any differences. Base your hypothesis on whether you think the very apparent differences between cows and chickens mean that they must have very different DNA.

> ### Scientific Surprise
>
> This experiment is a little different from most in this book in that there are no actual controls and variables. The experiment is simply the extraction and comparison of DNA.

# Materials You'll Need for This Project

As stated a bit earlier, this experiment is a little more complicated than most in this book, and requires some materials (such as liver) that aren't too likely to be lying around the house.

Make sure you'll be able to get the materials you need for the experiment before you commit to doing the project. They are listed below.

- ½ pound fresh cow's liver
- ½ pound fresh chicken liver
- meat tenderizer or enzymatic contact lens cleaner
- Kitchen knife
- Blender
- ½ teaspoon (2.5 ml) salt
- Two tablespoons (31.0 ml) dishwashing liquid (Dawn)

- One cup (354.0 ml) warm water

- Narrow glass or small test tube

- Beaker or large drinking glass

- Cheesecloth or coffee filter

- One cup (354.0 ml) cold isopropyl alcohol (rubbing alcohol)

- Glass, wooden, or plastic stirring rod

- A microscope

Once you've gathered all the materials, you'll be ready to begin the project.

# Conducting Your Experiment

Because you'll be using a knife and blender for this experiment, you'll need to work carefully. Be sure that you have adult approval and supervision before you begin working.

Follow these steps:

1. Using the kitchen knife, cut up the cow's liver into three small pieces measuring approximately one-inch (3 cm) cubed.

2. Place the pieces of liver into the blender.

3. Add the salt and warm water to the liver in the blender. Make sure the water covers the liver pieces.

4. Blend on high with the cover on for 10 to 15 seconds.

5. Strain the blended mixture through cheesecloth or a coffee filter into a beaker or large drinking glass.

6. Add the dishwashing liquid, stir, and let sit for 10 minutes.

7. Fill a test tube or small narrow glass one third full with the liver mixture.

8. Add a pinch of either the meat tenderizer or three drops of the enzymatic contact lens cleaner. Wait 20 minutes.

9. Tilt the test tube or glass, and slowly add an equal amount of the cold alcohol by pouring the alcohol down the sides of the test tube. Wait 20 minutes.

10. Stringy DNA should appear as a layer between the bottom mixture and the alcohol on the top.

11. Insert a stirring rod (either glass, wooden or plastic), into the stringy DNA. Spool the DNA around the rod like you are gently swirling up cotton candy.

12. Observe a smear of the DNA under a microscope. You also can dry the DNA on paper towel to further observe it and its characteristics.

13. Repeat the entire process with the chicken liver.

**Standard Procedure** _____

DNA will keep indefinitely when covered with alcohol in a sealed container. If you do this experiment at home and don't have a microscope, you could carefully transport it to school to observe it under a microscope.

# Keeping Track of Your Experiment

The quantitative aspects of this experiment are in the procedural steps. You must follow all the steps of the experiment carefully, keeping track of what you've done.

Your analysis of the extracted DNA, however, will require more qualitative work. You'll need to describe what you observe, including the DNA's color, texture, and general appearance. Write down all of your observations.

Studying the DNA under the microscope will provide you with a greater, more detailed view of this nucleic acid as a stringy structure.

**Standard Procedure** _____

Quantitative results are those that can be measured. Qualitative results are those that can only be observed.

# Putting It All Together

Record all your observations carefully, taking particular care to note the differences between the cow and chicken DNA.

Because your results are primarily qualitative, it's very important that you make careful observations and keep detailed notes.

# Further Investigation

If you're really into DNA after this experiment, there are all kinds of additional investigations you can pursue. You could further this experiment by extracting DNA from other sources, such as bananas, kiwi fruit, and animal muscle and compare the resulting DNA.

Remember that all living organisms—not just animals—have DNA. It would be fun to see how the DNA of a tomato, for instance, compares with that of a pineapple.

# Other Great Biology Projects

As you know, there's a lot more to biology than DNA. If you enjoy working in this field, you may want to try one or both of the experiments described below.

## Do Plants Sweat Like People Do?

You know that plants need water to survive, just like people and animals do.

But did you ever wonder whether plants sweat, like people do? Well, they don't sweat, exactly, but they do something similar. It's called *transpiration*, and it's when plants release water through their leaves.

This experiment isn't very difficult, and it's a good way to see some transpiration, firsthand.

**Basic Elements**

Transpiration is a process that occurs in plants when they release unnecessary water through their leaves.

What you need to do is break off a piece of a healthy geranium plant. If you're doing this at home, make sure it's okay with the adult in your house. You'll only need a piece of stem and one or two leaves.

Cut out a rectangle about four by six inches from a piece of thin cardboard, such as a shoebox lid. Poke a hole in the middle of the rectangle that's just big enough for the geranium stem to poke through. Poke the bottom stem through the hole.

Fill a water glass about three-quarters full. Put the cardboard rectangle on top of the water glass, with the stem in the water and the leaf on top of the cardboard. Rub a small amount of Vaseline around the hole. That seals the hole and prevents evaporated water from escaping.

Get a glass that's the same size and type as the first one, and place it upside down over the geranium leaf onto the cardboard. The mouth of one glass should match up with that of the other.

Put the glasses on a bright windowsill and let them sit for at least four hours. Then check out what's happened inside the top glass. You should see little drops of water on the inside of the glass. You know that the water didn't travel up through the hole in the cardboard, because you sealed that off. Look at the underside of the geranium leaf. The little dots you'll see there are called stomates, and they are the source for this water. The water passes through the stem, into the leaf, and what isn't needed is released through the stomates.

## What Types of Bacteria Are Found in Your House?

Bacterial organisms are everywhere. They are in our homes and schools, in our foods, and in the air we breathe. Your body is marvelously designed to fight bacteria when it enters your body through a cut or another means. While some bacteria is necessary in your body, too much, or the wrong types, can cause serious problems.

Bacteria are smaller than cells and often invade cells. If your body can't protect itself through its immune system, the bacteria will eventually destroy the cell.

One of the problems with bacteria is that you can't see it. If you could, you could do a better job of avoiding it. In the experiment described below, however, you get to not only collect bacteria, but see it grow, as well.

You'll collect samples of bacteria from different areas in your home, then allow it to grow in covered petri dishes containing the nutrient agar. Petri dishes are available from biological supply companies (see the listings in Appendix C). A set of 10 dishes will cost you about $15. Be sure to label each dish carefully so you know from which area the bacteria inside of it was taken.

**CAUTION**

**Explosion Ahead**

Much caution is needed while doing this experiment because you don't want the bacteria you collect to spread anywhere except in the petri dish in which you'll be placing it. Allowing bacteria to spread is not a good idea.

Some good areas within your home to check are door handles, the insides of your sneakers, staircase railings, bathroom faucets, your teeth, and the computer's keyboard.

Use a cotton swab to wipe the surface you're testing, then wipe the swab gently onto the nutrient gel in the petri dish. Put the cover, also labeled, onto the petri dish and place it in a dark, warm area where the bacteria can grow undisturbed.

Check for bacterial growth every 12 hours for one week. Record any changes in each dish noting the amount, color and shape of the bacteria.

At the end of a week you will be able to clearly see what areas in your home have the most bacteria. You can chart your information, providing helpful information for your parent or guardian.

Make sure you throw away the petri dishes at the end of the experiment. It would be a good idea to photograph each dish at the end of the experiment for your display at the science fair.

An interesting step further would be to test the back doorknob, then wipe it clean with bleach and swab it again and test for bacteria. This would help you to see if the cleaner did its job and killed the germs.

## The Least You Need to Know

- ◆ DNA, necessary for life in all living organisms, is short for deoxyribonucleic acid.

- ◆ DNA contains a chemical code that's used to make proteins in various cells, including muscle, blood, skin, bone, and nerves.

- ◆ The experiment in this chapter calls for you to extract different types of DNA from its nucleus and then compare it to see any differences.

- ◆ Chicken and cow livers contain many cells, and in turn, also DNA and proteins.

- ◆ Your observations of the extracted DNA will be largely qualitative, not quantitative.

- ◆ DNA can be extracted from all sorts of plants as well as animal tissue.

# 18

# Can Cloning Make a Better Plant (and Other Great Botany Projects)?

## In This Chapter

- ◆ Different types of cloning
- ◆ Understanding horticultural cloning
- ◆ Various methods of cloning a plant
- ◆ Comparing stem-cutting plants with the parent plant
- ◆ How distance affects the effectiveness of pesticides
- ◆ Understanding the role of microorganisms

We read a lot about cloning in the news these days. Ever since Scottish scientists created "Dolly," the first cloned sheep, in 1997, there has been lively debate about cloning and the possibilities it raises.

There are several types of cloning, although we tend to think primarily of cloning in the reproductive, or Dolly sense of the word. In Dolly's case, scientists removed the nucleus of an egg cell taken from one sheep, and injected a mammary gland cell from another sheep into the nucleus-free egg cell. This specialized cell, which in time became Dolly, was implanted in the uterus of a third sheep, who served as Dolly's surrogate mother.

In addition to reproductive cloning, molecular biologists have been cloning genes and DNA since the 1970s for laboratory work. And therapeutic cloning (sometimes called embryo cloning) research is under way, with the goal of harvesting stem cells that can be used to treat disease and for other purposes.

Therapeutic cloning has drawn a great deal of criticism and generated ethical concerns because it requires the use of human embryos.

There are plenty of books and websites from which you can learn more about all types of cloning. If you're interested, it would be a good idea to learn what you can, as cloning promises to continue to be a hot topic in our society.

For the purposes of this experiment, however, you needn't get excited about stem cells or DNA or anything like that. The experiment in this chapter deals with plant cloning, which is simply making a plant that's identical to another.

# So What Seems to Be the Problem?

The problem you'll attempt to solve during the course of this science fair project is whether a plant you clone from another plant is better than the original. Let's explain what that means.

In horticultural terms, the word "clone" is used to designate all the descendants of a single plant, produced by vegetative methods. This means that the plants multiply by means other than seeds.

| Scientific Surprise |
| --- |
| A plant produced using the vegetative method is said to have been asexually reproduced. The production of seeds for the purpose of producing plants is called sexual reproduction. |

All of the plants that are produced from one plant have a common origin, and are considered extensions of a single plant.

Cloning of plants, or artificial plant propagation, commonly involves separating a portion of the parent plant in order to produce an independent plant. Almost any part of the plant, including roots, stems, and leaves, is capable of vegetative reproduction.

The cloning of a plant can be accomplished in several ways, including the following:

♦ **Leaf cuttings.** Removing a leaf from the original plant and planting the stem of it in soil can eventually produce a new plant. This occurs with common plants such as African violets, snake plants, gloxinia, and several species of begonias.

Certain plant leaves, when cut from their stems and placed flat on top of the soil, develop roots at particular intervals along the cut veins of the leaf. Eventually, the parent leaf looks as if it has sprouted several baby plants, all growing vertically from the leaf at various points along the veins. Plants that reproduce in this manner include begonia and geranium.

♦ **Stem cuttings.** Removing a piece of a stem from a plant and inserting it in soil either horizontally or vertically is another method of cloning. Eventually, roots will develop downward from the stem, and a new plant will emerge from the top. When a bud is emerging from a stem and the stem is cut from the plant, a new plant will develop from the bud. Plants capable of this sort of reproduction include begonia, gardenia, Christmas cactus, lantana, and impatiens.

♦ **Budding.** This method of cloning is common in the propagation of many fruit trees. A bud, which is an undeveloped branch, leaf or flower protruding from the stem of the plant, is cut from the parent, placed into a notch made into the parent stem, wrapped in place and allowed to grow until it can be removed from the parent and allowed to grow in soil on its own.

♦ **Plant division.** Many flowers, such as the commonly found daylily, African violets, certain orchids, and ferns, are cultivated through the division of their thick roots. Once divided, each part can be potted and grown as a new plant.

♦ **Runners.** Wild strawberry plants are among those that put out horizontal stems that touch the soil. These stems can eventually develop roots from which another strawberry plant will grow. The new plant is identical to the parent.

♦ **Grafting.** A cutting from a plant is attached to a piece of the root, or to the rooted stem of another plant. The two pieces become united, and will grow as one plant. In this method, desirable properties of one plant can be mated with the desirable properties of another to produce a new *hybrid* plant without producing hybrid seeds. Producing dwarf fruit and oriental trees for home landscaping is a common use of the grafting method.

**Basic Elements**

A **hybrid** plant is the offspring of two varieties or species of plants. A tangelo, which is a mixture of an orange and a tangerine, is an example of a hybrid.

So as you now know, there are numerous ways to clone plants. Our question is whether the cloned plant is stronger than the original. Does it live longer? Grow better? Resist disease better?

If you want to, you can use the name of this chapter, "Can Cloning Make a Better Plant?" as the title of your science fair project. Or, you might want to consider one of the titles listed below.

◆ How Does Cloning Affect the Quality of a Plant?

◆ To Clone or Not to Clone

If none of those titles strike your fancy, go ahead and make up one of your own. Let's get started.

# What's the Point?

Cloning allows for the possible propagation of large numbers of plants with desirable traits, such as a high fruit yield or resistance to disease. When plants are propagated, it means that they're produced in large numbers in order to create a supply of them.

| **Scientific Surprise** |
|---|
| Dolly, the sheep cloned by Scottish scientists in 1997, was put to sleep in February 2003. The average age for Dolly's breed of sheep is about 12 years, but Dolly suffered from cancer and other ailments. |

In the world of fruit and vegetable farming, having high-producing plants and trees is advantageous, and often necessary for economic survival. A farmer who can get 50 tomatoes from one plant will be better off than one who's only getting 30 tomatoes.

Most plants employ a method called self-fertilization in order to reproduce. Seeds are produced during this self-fertilization process. Those seeds then fall to the ground, or are spread by wind or birds or other means.

Some plants cross-fertilize—or cross-pollinate—to produce their seeds.

Insects often carry the pollen, which are the male eggs, from the flower of one plant to the carpal, which is the female receptacle, of a flower of another plant. Seeds form as a result. The plants that grow from these seeds are a cross between the two different plants.

If you want an exact copy of a plant, however, without waiting for self-pollination to occur, then cloning is the way to go.

In the experiment outlined later in this chapter, you'll use a method of vegetative reproduction (stem cuttings) to clone plants from a parent plant. Your control will be the parent plant, and the cloned plants will be your variables.

# What Do You Think Will Happen?

Stem cuttings are a common, and fairly easy, method of achieving vegetative reproduction. You will need a begonia plant, which is a very common plant variety. If you can't find a begonia, you can try the experiment using another type of plant. A begonia, however, should not be difficult to locate.

After you've taken cuttings from the parent plant and grown them into separate plants, you'll want to compare the parent plant with the results of the vegetative reproduction you've achieved. You'll want to see which of the plants appear stronger, have brighter flowers and foliage, produce more flowers, and so forth.

Take a moment here to make a guess about how the plants will compare. Do you think the older plant will be healthier because it's more established? Or that the newly rooted plants will be because of their new growth? Perhaps taking root cuttings from the parent plant will stress the plant to the point that it will become unhealthy.

Go ahead and venture a hypothesis about how you think the cloned plants will compare to the parent plant, and then you'll learn how to take a root cutting and what to do with it once you've taken it.

# Materials You'll Need for This Project

You will need, as stated earlier, a good-sized begonia plant, from which you'll take five stem cuttings. It's best to use a sharp, serrated knife (like a steak knife), to cut a stem from a plant. Scissors or a nonserrated knife tend to crush the plant tissue.

To take a stem cutting, make a diagonal cut on a main stem or a side stem. The stem pieces you cut should be about two to three inches long, and contain two leaves at two different points along the cut stem.

To better ensure that roots will form, you can dip the stem into a rooting enhancer such as Roottone or Hormonex. These products contain a chemical that helps generate roots. Using such a product for this experiment is recommended, but not necessary.

*You'll need five stem cuttings for your experiment.*

Materials you'll need for this experiment are listed below.

- One begonia plant
- Sharp, serrated edge kitchen knife
- Rooting enhancer (preferred, but not necessary)
- Small bag of potting soil
- A plastic, shoebox-size container in which to initially grow the stems
- Plastic wrap or a piece of glass to cover the containers
- Plastic pots in which to transplant the growing plants later in the experiment
- Metric ruler

**Standard Procedure**

This project requires a significant amount of time to complete properly. It could take you up to four months to grow the stem cuttings and have adequate time to compare them with the parent plant. Be sure to begin working on the experiment in plenty of time so that your project can be completed before the fair.

Once you understand how to make root cuttings and have assembled the necessary materials, you're ready to begin the experiment, as outlined next.

# Conducting Your Experiment

Be sure to have everything you need ready to go before you begin the experiment. The stem cuttings should be planted into the soil soon after they're made.

Follow these steps:

1. Pour the potting soil into the container in which you'll be planting the stem cuttings.

2. Make five stem cuttings from your begonia plant, cutting diagonally along the stem.

3. Dip the cut ends of the stems into the root enhancer, if using it.

**CAUTION** **Explosion Ahead**

The root enhancer is not necessary, but if you use it, be sure that you use it on all the stem cuttings, not just some. Treating only some of the cuttings would create another variable in the experiment.

4. Place the stem either vertically or horizontally in the soil. If planting the stem upright into the soil, insert about one-third of the stem into the soil. If you lay the stem in the soil horizontally, bury the stem to a depth equal to twice its width.

5. Moisten the soil by misting it with water from a spray bottle.

6. Cover the container with either plastic wrap or a piece of glass.

7. Place the container in a warm, bright area but not in direct sunlight.

8. Mist the soil every other day.

9. Check the stem cuttings at regular intervals (weekly would be ideal), recording any changes you notice as the stems begin to grow within the container. Be sure to measure and record the height of the plant, and to note other changes and characteristics that you observe.

**Standard Procedure**

Providing warm (room temperature) soil and high humidity will greatly help along the propagation process. This can be accomplished by covering the planting container with plastic wrap or a piece of glass to simulate a greenhouse. Do not place the cuttings in direct sunlight until they are well established.

10. Once the stems are established and growing, transplant them into individual pots. Continue to observe the new plants—and the parent plant—over a period of two or three months, keeping close track of what you see.

*Plant the established cuttings in individual pots.*

# Keeping Track of Your Experiment

Use the following charts, or make your own charts, to record growth rate and other observations about the five stem cuttings and the parent plant.

Notice which plants appear healthiest, have the most blooms, and so forth. Be sure that all the plants are kept under the same conditions (amount of light and water, and so forth) until your experiment is complete.

*While you're recording the growth rate and other features of the new plants, be sure to continue observing the parent plant. Record what you see happening to that plant, as well.*

**Observations and Data Chart #1**

| Stem # and Date of Observation | Length of Stem (in cm) | Number of Buds on Stem | Number of Leaves on Stem |
|---|---|---|---|
|  |  |  |  |
|  |  |  |  |
|  |  |  |  |
|  |  |  |  |
|  |  |  |  |

**Observations and Data Chart #2**

| Date (Record) | General Observations on Growth of New Plant #1 |
|---|---|
|  |  |
|  |  |
|  |  |
|  |  |
|  |  |
|  |  |
|  |  |
|  |  |

*Copy this chart, or create a separate chart for each of the five stems, and use them to record general observations during the weeks of new growth.*

# Putting It All Together

When you've completed the experiment and have had enough time to observe growth and changes in both the root cutting plants and the parent plant, consider what you've seen.

Do the new plants appear healthy? Are the leaves and foliage of the new plants the same colors as the parent plant? How do the new plants compare in size to the parent plant?

So what have you learned? Does cloning make a better plant or simply an identical plant?

# Further Investigation

Many plants that can be propagated through stem cuttings can be rooted in water instead of soil. This method does not require creating the additional humidity, as you did when you rooted cuttings in dirt.

If you're interested in taking the experiment you completed a step or two further, you could root some cuttings in water, and then compare the growth rates and success rates of the two methods.

Begonias can also be propagated by removing a leaf, cutting the veins of the leaf, and placing the leaf on top of warm soil in a humid environment. Place small stones on top of the leaf to make it lie flat. New plants will grow at the places where the veins were cut.

# Other Great Botany Projects

Botany is a fascinating area of science with many opportunities for practical application. As you probably know, agricultural methods have become increasingly sophisticated over the past decades, depending largely on science, including botany.

If you're interested in botany, you could follow up on the projects suggested below. One involves the use of pesticides, while the other explores the possibility of growing plants in soil that contains no microorganisms.

## How Close to a Plant Must Pesticides Be in Order to Be Beneficial?

There often is controversy over the use of pesticides. And yet, many people don't have a good understanding about exactly what pesticides are, and why they are often necessary for plant production.

A pesticide is anything that destroys pests or suppresses or alters their life cycles. We normally think of pests as bugs, but they also can be invasive plants, fungi, bacteria, and other organisms.

Pesticides are not necessarily synthetically produced. Many pesticides occur naturally. Pesticides are used to control bacteria, fungi, and insects; to control rodents and other animal pests; to control pest-category plants; and to repel pests.

In this suggested science fair project, you'll seek to discover how close to a plant a pesticide must be in order for it to be effective. You could choose any pesticide that you wish, but you'll need to use the same type of plants, growing conditions, and so forth.

You could do this experiment with potted plants set outside, but it would work better if you could plant four groups of the same plant at short distances apart directly in the soil.

Apply pesticide directly on one group of plants, according to manufacturer's instructions. Then, place pesticide near, but not on, the other groups of plants, varying the distance from group to group.

Observe how the different groups of plant fare as far as insect damage is concerned, in addition to their general health.

## Can Healthy Plants Be Grown in Soil Containing No Microorganisms?

Soil is home to a great number and wide range of microorganisms, among them algae, fungi, bacteria, and viruses.

Although the number of microorganisms found in soil can sound alarming, it's important to know that these tiny organisms need nutrients in order to grow and be active in the soil. If the soil lacks the necessary nutrients, the majority of microorganisms may be physiologically inactive.

Microorganisms can be both good news and bad news for plants and those who grow them. Some help to decompose roots, stems, and leaves of plants that die, helping to recycle nutrients in the soil. Others increase nutrients in the soil and make it more fertile. Some even act as natural fertilizers.

> **Scientific Surprise**
>
> There normally are between one and 10 million microorganisms present in one gram of soil—which is only $\frac{1}{28}$ ounce—a tiny, tiny amount. There are more bacteria and fungi than any other type of microorganisms.

Not all microorganisms are helpful, however. Some can infect plants through their roots and cause considerable damage.

In this suggested science fair project, you would investigate whether plants grow better in soil containing microorganisms than in soil that has none. To do so, you'd take a sample of fertile soil from your garden or a field, and divide it into two parts. Bake one part in the oven at 350 degrees for one hour to destroy all the microorganisms, and leave the other batch alone.

Plant the same number of the same type of seeds in each of the different batches of soil, and provide the same environment and care to each group. Record your results carefully.

# The Least You Need to Know

◆ Three main types of cloning are reproductive cloning, genetic cloning, and therapeutic cloning.

◆ Cloning has been around for years, but continues to be a controversial topic.

◆ There are various methods, including stem cuttings, of achieving vegetative reproduction.

◆ Stem cuttings must be kept in a warm, humid place in order to ensure successful rooting.

◆ Stem cuttings can be compared with the parent plant after they've become firmly established and transplanted into their own containers.

◆ Pesticides and microorganisms are other topics relevant to the area of botany.

# Chapter 19

# Which Metal Corrodes the Fastest (and Other Great Chemistry Projects)?

## In This Chapter

- ◆ The basics of rust, oxidation, and corrosion
- ◆ The importance of different types of metals
- ◆ Using your own experiences to formulate a hypothesis
- ◆ Knowing what materials you'll need
- ◆ Testing the different metals
- ◆ Other projects to consider

Did you ever have a shiny new bike that over time got to look not so shiny and new anymore? Or some beach chairs that got left outside over the winter and by spring looked like they were ready for the trash heap?

If so, chances are that the culprit was rust. Mailboxes, swing sets, lamps, cars, railings, and just about anything else made from metal are subject to the perils of rusting—a deteriorating metal condition.

You've undoubtedly had some experience with rust—or at least have seen it on a car or other object. Rust is so common that its color also is called rust, as in rust-colored leaves, or rusty-brown hair.

In this chapter, you'll learn a lot more about rust and how it occurs. And you'll attempt to find out which metals rust the fastest, by exposing them to water and to salt water.

# So What Seems to Be the Problem?

Rust occurs when metals containing iron react with the oxygen in the air or in water and form a compound called iron(III) oxide (ferric oxide). This compound contains water molecules, so we call it a hydrated compound.

Both oxygen gas and water must be present for the iron to rust. Chemically and very simply speaking, iron atoms lose a few electrons to oxygen atoms. This process by which electrons are removed from atoms is called *oxidation*. When oxidation occurs, it produces a chemical reaction that creates iron(III) oxide—or rust.

Rust is a type of corrosion. But it's not the only type. Other forms of corrosion include:

♦ Tarnish found on silver teapots, trays, and jewelry

♦ Copper carbonate, or patina, the corrosion that causes copper to turn green

♦ Discolored spots that appear on brass

♦ Aluminum oxide, which forms on aluminum

♦ Chromium oxide, which forms on the outside layer of stainless steel

**Basic Elements**

Oxidation is the process by which electrons are removed from atoms. It also can refer to a reaction of an object exposed to oxygen.

On some metals, corrosion actually serves as a type of protection. Aluminum oxide, copper carbonate and chromium oxide, for instance, act as protective coatings for the underlying metals.

Rust that forms on iron, however, cannot protect the iron from further corrosion because it's too porous.

The problem you'll be attempting to solve in this science fair project is which metals corrode the fastest, and under which conditions. You'll test five metals—silver, steel, zinc, copper, and aluminum—to see which corrodes fastest in water and in salt water.

When you've finished, you'll have a better understanding of corrosion, the process of oxidation, and the properties of different metals.

The title of this chapter, "Which Metal Corrodes the Fastest?" would be a suitable title for your science fair project. Other possible titles include:

- ◆ Which Metal Holds Up Best in a Corrosive Environment?

- ◆ Understanding How Corrosion Affects Common Metals

> **Scientific Surprise**
>
> Corrosion causes tremendous damage to buildings, cars, bridges, and ships. Finding a method to halt corrosion is a high priority for experts working in the metal industry.

Or, you can think of a name for your project on your own. Let's take a few minutes now to consider why this project is valuable.

# What's the Point?

Why should you care which of the five metals you'll be testing corrodes the quickest? Why should you care about metals at all, for that matter?

Metals have thousands of uses that affect our everyday lives, most of which we take for granted. Copper, for instance, is pliable and a good conductor of electricity. For those reasons, it's used to make the wire inside of electrical cables. Without electrical cables we'd have no electricity in our homes—no light, TV, or video games.

Aluminum is extremely strong and can be fashioned into thin sheets, making it vital for aircraft production. Think about that the next time you climb onto an airplane. Metals are used to make the utensils we eat with, the coins we use to buy what we want, and the cars we drive.

Obviously, metals that are used to build aircraft, cars, and electrical wiring had to be extensively tested to make sure they were suitable for use.

You can be sure that there was far-reaching research and experimentation before the first copper wire was put to use in an electrical cable. *Metallurgists*—experts on metals—are constantly looking for new uses of metals in many fields, including medical, military, and aeronautics.

Metals—and how they're used—are extremely important. Once you know how different metals

> **Basic Elements**
>
> A **metallurgist**, sometimes called a metallurgical engineer, researches, controls, and develops processes used in extracting metals from their ores in order to refine them. Experts in the area of metals, metallurgists also study the effects of combining metals with other materials, such as polymers and ceramics.

hold up to corrosion, you'll be able to better understand why they have particular roles, and why they're important. In addition, you'll be able to give Mom and Dad some pointers the next time they're shopping for a new outdoor lamp or metal toy for your little brother.

By experimenting with five different types of metal wire, you'll be able to see which corrodes the fastest, and which ones hold up best under certain circumstances. You'll test each wire in both distilled water and salt water. Again, the types of metal you'll be testing are:

- ◆ Silver
- ◆ Steel
- ◆ Zinc
- ◆ Copper
- ◆ Aluminum

Your control group will be 10 pieces of wire—two each of the metals listed above. The variables are the distilled water and the salt water in which the metal wires will be immersed. Using the scientific method, you'll learn which metal begins to corrode first, and which holds up the best.

# What Do You Think Will Happen?

Think about what you may already know about different kinds of metals and how they react when exposed to rain, or air or water that contains a lot of salt. This will help you to formulate a hypothesis based within the context of knowledge you already possess.

Go back to the bicycle mentioned in the first sentence of this chapter. Under what type of circumstances did your bike rust? When it was stored in the dry garage? Or when you left it lying out in the yard for three days during a steady rain?

Why do you suppose that cold-weather drivers are advised to rinse off their cars every now and then during the winter season when road salts are being used? Have you ever noticed or heard people talk about problems with corrosion near the beach, where salt water is prevalent?

Do you already know, perhaps, which metals are most resistant to corrosion? If so, the experiment you'll do will support and affirm your knowledge. If you don't, try to use common sense and any information you may have about this topic to come up with your best guess—or hypothesis.

# Materials You'll Need for This Project

The experiment you'll be doing will require only a short amount of time to set up, but you'll need to make observations over a 10-day period.

It's going to be important to write down exactly what you see happening to each metal each day. Remember that your measurements will be qualitative, not quantitative. For that reason, the more data you present concerning your experiment, the more reliable your results will be.

You'll need some materials for this experiment that probably aren't lying around your house. You should be able to find everything you need, however, at your local hardware or home supply store. You will need:

◆ 12 inches (30.5 cm) of silver wire

◆ 12 inches (30.5 cm) of steel wire

◆ 12 inches (30.5 cm) of zinc wire

◆ 12 inches (30.5 cm) of copper wire

◆ 12 inches (30.5 cm) of aluminum wire

◆ Small pair of wire cutters (or ask the person at the store who cuts the wires for you to cut each 12-inch piece in half)

◆ 10 clear drinking glasses (preferably identical), or 10 test tubes and a test tube rack

◆ A pen or fine-point marker

◆ Small pieces of paper or labels for the glasses or test tubes

◆ 10 pencils (they don't need to be sharpened)

◆ Transparent or masking tape

◆ Liquid measuring cup

◆ A tablespoon measure

◆ A funnel

◆ Distilled water (available in gallon jugs in most grocery stores)

◆ Table salt

If you can get test tubes and a rack, you'll probably find them easier to use than glasses. If you have to use glasses, however, that's fine. You can use plastic or glass cups; just make sure that they're clear so you're able to easily observe what's happening to the wires in them.

# Conducting Your Experiment

Make sure that you have all your materials ready before you begin the experiment. Be sure to find an area large enough to accommodate the glasses or test tubes, where they will be undisturbed for the duration of your experiment.

Follow these steps:

1. If the five wires aren't already cut, cut them into 6-inch lengths.

2. Using the pen or marker, mark ten labels or small pieces of paper as follows:

   water + silver            salt water + silver

   water + steel             salt water + steel

   water + zinc              salt water + zinc

   water + copper          salt water + copper

   water + aluminum       salt water + aluminum

3. Set the glasses or test tubes on a table or counter where you'll be able to easily observe them.

4. Stick a marked label, or tape a piece of marked paper, on each glass or test tube. Face all labels front so you can easily see them.

5. Using the measuring cup and funnel, fill five glasses or test tubes with distilled water.

6. Mix 8 ounces (240 ml) of water with 1 tablespoon of salt. Stir until the salt is completely dissolved.

7. Fill the other five glasses or test tubes with the salt water solution, mixing more water and salt as needed.

8. Wrap one end of each piece of wire around a pencil, so that when the pencil rests across the top of the glass, the wire hangs to the bottom.

9. Observe each wire at least once a day for 10 days. Use the charts found in the following sections, or make your own charts.

*Wires of different materials are suspended in distilled water and salt water.*

Remember that the more clear and accurate your observations are, the better you'll be able to draw conclusions from your experiment.

## Keeping Track of Your Experiment

Use the charts on the following pages, or make your own, similar charts to keep track of what you observe during the course of your experiment.

Be sure to not mix up the glasses. They'll all look very similar, so be sure that the labels remain intact and you can see them clearly.

**CAUTION**

**Explosion Ahead**

Don't be tempted to cut short your observation time, even if one or more of the wires appears corroded before the 10-day period has ended. Decreasing the experiment time will jeopardize the reliability and validity of your results.

## Putting It All Together

Some observations you'll want to consider are how the changes to the metal wires immersed in the distilled water compared to the wires in the salt water. Which metals had the most rust? Was the formation of the rust on any of the wires concentrated on one particular area on the wire? Or was the corrosion distributed evenly along the immersed wire? Based on your data, which metal would you recommend for the manufacture of bikes, beach chairs, and swing sets—not to mention aircraft and medical equipment?

Once you've recorded your results, you can draw a conclusion and identify the answer to the problem you stated at the beginning of your project.

**Observations of Wire Metals Suspended in Distilled Water**

| Day | Silver | Steel | Zinc | Copper | Aluminum |
|-----|--------|-------|------|--------|----------|
| 1 | | | | | |
| 2 | | | | | |
| 3 | | | | | |
| 4 | | | | | |
| 5 | | | | | |
| 6 | | | | | |
| 7 | | | | | |
| 8 | | | | | |
| 9 | | | | | |
| 10 | | | | | |

**Observations of Wire Metals Suspended in Salt Water**

| Day | Silver | Steel | Zinc | Copper | Aluminum |
|-----|--------|-------|------|--------|----------|
| 1 | | | | | |
| 2 | | | | | |
| 3 | | | | | |
| 4 | | | | | |
| 5 | | | | | |
| 6 | | | | | |
| 7 | | | | | |
| 8 | | | | | |
| 9 | | | | | |
| 10 | | | | | |

# Further Investigation

If you enjoyed this project and would like to take it a step or two further, you could try one of the following ideas:

◆ Place the metal wires in different liquids and see what happens. You could try vinegar, club soda, coffee, tea, soy sauce, or any other nonhazardous substance.

◆ Try using different metals, such as brass, titanium, or zinc.

◆ Test to see whether different conditions lead to different results. If you place some of the glasses in a cold spot, for instance, and others where it is very warm, do you get different results between the two groups?

Use your imagination to come up with other ways to vary the project and delve a bit further into this issue. Just be sure to keep good, accurate notes.

**Standard Procedure** _____

A good idea when presenting your project would be to display any corroded wires next to a new piece of the same type of wire. If you want to do this, remember to buy an extra 6 inches of each type of wire so that you'll have a new piece at the end of the experiment.

# Other Great Chemistry Projects

If chemistry is a field you're interested in, you can be assured that there are many, many projects waiting to be completed. Two such projects are described below, although not in as much detail as "Which Metal Corrodes the Fastest?"

## What's in a Color?

Are all colors created equally? Is the red dye in your favorite magic marker simply made up of red chemicals? Is the black marker composed of just black dye? What about your blue marker? Let's find out.

The process of separating the colored compounds in a mixture is called *chromatography*. The process can be a simple one as in the experiment suggested. Or, the separating procedure can be quite involved. You can conduct an experiment that actually separates the colors in magic markers into different colored layers on small strips of paper towels.

**Basic Elements** _____

**Chromatography** is the process of separating the color compounds found in a mixture. If you separate black, for instance, you'll discover red, yellow, and blue.

The colors that make up the inks found in magic markers have a physical property that allows them to move individually on a vertical basis at different rates. The different molecules of the mixture move up the paper towel as they travel with the liquid solvent as it is being absorbed.

In this experiment, you'll be able to watch bands of different colors appear on a paper towel or chromatography paper as the ink in the marker separates and travels along the paper, which is suspended in water. Remember, before you begin a science project, research your topic so you have a better understanding of your subject. Normally, the more you know, the easier it will be to hypothesize, conduct your experiment, and draw correct conclusions from your results.

You'll need a paper towel, or chromatography paper, if it's available. Chromatography paper is a narrow paper (about 2 centimeters wide) that comes on a roll. It can be purchased at a science supply store. You'll also need scissors, a blue nonpermanent magic marker, a rubber band or piece of tape, a small container such as a plastic or glass cup, and some water.

Cut strips of paper towel approximately 12.0 cm by 3.0 cm, then cut the bottom of each strip on an angle, so that the bottom is a point.

*Use a paper towel or chro-matography paper cut like this for your experiment.*

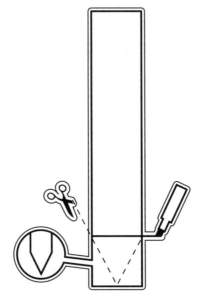

Use a magic marker to draw a line across the bottom of the strip, above the point. Half fill the cup with water. Hang the strip with the tip just in the water. Bend the top of the paper towel over the rim of the glass and secure it with a rubber band. Record your observations after 24 hours.

By experimenting with different colors of markers, you can compare their color components. You also could vary the experiment using just one type of marker, but changing the separation solvent. Use water as a control, and other liquids such as white vinegar, rubbing alcohol (isopropyl alcohol), salt water, fingernail polish remover (acetone), and ammonia. Which liquid separates the colors of your markers the best? On a practical note, you may discover a solvent that will remove that magic marker stain from your new khakis.

## Copper Magic

This is an easy experiment to conduct, but interesting because it involves several chemical reactions and you get some pretty dramatic results.

In this science fair project, you transform dull pennies into shiny ones, then watch them turn bluish-green. At the same time, you can watch rust bubble off of steel nails.

You may have to do some further research to fully understand this project. The basics of the experiment, however, are outlined below.

Place 20 pennies (the dirtier and uglier the better) in a shallow glass bowl containing 1 teaspoon (5.0 ml) of salt, mixed with a quarter-cup (50 ml) white vinegar.

Presto! You will witness a chemical reaction almost immediately. The tarnish on the pennies is called copper oxide and it will react with the weak acid solution of salt and vinegar.

Leave the pennies in the bowl for about five minutes, then remove half of them. Rinse and dry them, and put them on a paper towel marked "rinsed." Then remove the rest of the pennies and, without rinsing or drying, let them dry on another piece of paper towel marked "not rinsed."

Add a steel screw and a steel nail to the bowl containing salt and vinegar. Lean another steel nail against the inside of the bowl so that part of the nail is in the salt solution, and the other part is not.

Wait for an hour, and closely look at all the pennies, the nails, and the screw. You should be able to record a lot of observations and data.

What happened to the pennies that were not rinsed and dried? These pennies actually

### Standard Procedure

A good display for this project would be to mount a dull penny next to a chemically altered shiny one. Also mount a greenish-blue penny to show how the coins changed throughout the experiment. You also could display a "before" rusty steel nail next to an "after" steel nail to illustrate the difference between them.

underwent two chemical reactions: one while they were immersed in the salt solution, and another as they dried on the paper towel.

You know that the reaction was caused by the acid solution. The other occurred when the atoms that make up the penny reacted with the oxygen in the air and the chlorine from the salt. Is there a color change on either of the paper towels? That bluish-green compound that formed is called malachite.

What about the steel screw and nails? How have they changed? What gaseous compound was formed and bubbled through the acidic solution? What would the balanced chemical equation be for the changes you saw?

If you change the concentration of the solution by varying the amount of dissolved salt in the vinegar, you are actually changing the pH of the solution. Test the acidity of each different solution you make using pH paper that tests liquids in an acid range. You usually can find pH paper at a hobby store that supplies science kits and equipment. Do you notice if a change in the pH increases or decreases the reaction rate? Would lemon juice or orange juice produce the same results? Try it and find out.

You can take this project a step further by discussing the chemical changes that occurred in terms of atomic and molecular levels.

## The Least You Need to Know

- Rust, oxidation, and corrosion are all factors that affect metals.
- Be sure to gather and organize all necessary materials before you begin the experiment, and think about issues such as space and convenience.
- Observe the metal wires for 10 days in order to assure adequate data.
- Varying the method in which you conduct the experiment could alter the results you get.
- Other projects involving chemistry include separating colors and transforming dull pennies into shiny ones.

# Chapter 20

# How Can Different Polymers Be Identified (and Other Great Earth Science Projects)?

## In This Chapter

◆ Learning how plastics can be identified

◆ Different types of plastics

◆ Testing the densities of different types of plastics

◆ Keeping close track of your observations

◆ Learning about the air you breathe

◆ Testing the acidity of snow and rain

It seems at times that we are a society drowning in plastics. We buy a package of cookies that's enclosed in plastic, then put it in a plastic bag to carry home.

People who don't recycle might open the cookies, stuff the plastic wrapping back into the plastic bag, and stuff that plastic bag into a plastic trash bag to be thrown away.

Fortunately, many people do recycle, and more and more are becoming aware of methods for cutting down on the amount of plastic they use. There are movements underway that encourage people to invest in reusable cloth bags for shopping, or to return plastic shopping bags to a recycling center at the grocery store.

Many communities offer recycling pickup, or have drop-off centers where residents can dispose of recyclable materials.

Recycling, however, does not happen all that easily. Getting your recyclables to the curb or to the drop-off center is the easy part.

Plastics are generally sorted by and recycled by type. There are seven types of plastic labeled for recycling purposes. The recycling code is normally on the bottom or side of a container, and is a number within a triangular set of three arrows. Separating and recycling plastics in this way produces a more consistent and higher-quality product.

Once separated, the plastic needs to be cleaned to remove contaminants. This normally is done by cutting the plastic into small pieces, which then undergo a washing and drying process. Those small pieces are then converted into plastic pellets, which can be used to make new plastic products.

Nearly everyone agrees that recycling is not only a good idea, but necessary if we are to continue to inhabit this earth. We can recycle materials other than plastic, of course, but because we use so much plastic, its recycling is of vital importance.

In this chapter, you'll learn more about how plastic is made, and the properties of different types of plastic. Your task, after working with different types of plastics for a while, is to see if you can identify each type without looking at the recycling symbol it contains.

# So What Seems to Be the Problem?

The problem is that recycling is a tedious process. Mostly, that's because the plastic needs to be separated by type. The problem you'll attempt to solve is whether you can differentiate between various types of plastic, based on their physical and chemical properties.

Once you're able to do so, you can appoint yourself as chief recycler in your household, or volunteer at your community's recycling center.

A plastic is a type of polymer. A polymer is a chemical compound that is made up of repeating units of molecules called monomers.

Monomers are made up of two or more carbon atoms, bonded to one another with hydrogen, which is bonded to the carbons. In some plastics, there can be other elements bonded to the carbon as well, such as chlorine or fluorine or nitrogen.

These different elements in the monomers contribute to the different properties of the different types of plastics.

If you want to, you can use the name of the title of this chapter, "How Can Different Polymers Be Identified?" as the title of your science fair project. Some other titles you might consider are listed here.

**Standard Procedure**

Physical properties of plastic include color, mass, volume, density, melting point, and the color of smoke produced when the plastic is burned. Chemical properties include whether or not the plastic is flammable, the color it burns, and its reactivity to other substances.

   ◆ How Do Your Plastics Stack Up?

   ◆ Physical and Chemical Properties of Different Types of Plastic

   ◆ Recycling Revisited

Or you can think of your own title for your project.

# What's the Point?

Knowing more about the plastics you use in your daily life will help you to understand their physical and chemical makeups, and be better able to appreciate the procedure of recycling.

In this experiment, you'll test the properties of six types of plastic. (You won't be working with any from the seventh category, which includes a wide range of substances under the grouping of "other plastics.") Because you'll be comparing all the different kinds, you won't have a control and variable. You won't be using one type of plastic as the control and comparing all the other types to it. You'll examine six types of plastic and make comparisons between all of them when you've finished.

As recycling technology improves and, hopefully, more attention is called to the importance of recycling, we will see additional methods of reusing plastics and other materials.

**Explosion Ahead**

Recycling plastic is definitely the way to go, but don't think of it as the ultimate answer to our trash problems. Most plastic items can be recycled, but in some cases, only once. Environmental activists urge consumers to use more glass and aluminum, which can be recycled repeatedly.

# What Do You Think Will Happen?

Once you have learned more about the properties of plastics, you can make an educated guess about how well you'll be able to identify different types.

You will be performing specific tests on six types of plastic, and taking notes about what you observe. The six types of plastic you'll need to work with are listed here.

1. **Polyethylene terephthalate (PET).** Nearly a quarter of all plastic bottles—including carbonate beverage bottles—are made from this type of plastic. So are meat wrappers, filling for pillows, and cosmetic wrappings.

2. **High-density polyethylene (HDPE).** This type of plastic is used to make about 60 percent of all plastic bottles, such as those containing milk, detergents, shampoo, bottled water, juices, and antifreeze. It's also used for plastic grocery bags and freezer bags.

3. **Polyvinyl chloride (PVC).** Some bottles—primarily those used to hold cleaning agents—are made from this type of plastic. So are electrical conduit, plumbing pipes, blister packs, and roof sheeting. A variation of this type of plastic is used to make garden hose, shoe soles, blood bags, and cable sheathing.

4. **Low-density polyethylene (LDPE).** This type of plastic is used in filmy-type bags such as garbage and bread bags, squeeze bottles, garbage cans, and irrigation tubing.

5. **Polypropylene (PP).** Microwave containers, drinking straws, potato chip bags, yogurt containers, plastic buckets, and plastic patio furniture all originate from this kind of plastic.

6. **Polystyrene (PS).** This kind of plastic is used to make deli and salad bar take-out containers, plastic cutlery, clear plastic cups, and plastic sleeves for cookies or crackers. A variation of this type of plastic is Styrofoam, used for beverage cups, meat-packaging trays, and protective packaging.

## Standard Procedure

Remember that we're not going to use Styrofoam, even though it is identified as a number 6 plastic. Get a clear, plastic drinking cup instead, or other item bearing the identification number "6."

If you take a look around your house, you can probably come up with a sample of each of the six types of plastic. Once you've rounded up a sample of each plastic 1 through 6 (raid your neighbor's recycling can if you need to), take a few minutes to look at what you've got. Notice the differences in the plastics. Some are smooth and pliable, while others are brittle and rigid. Some are clear, while others are opaque.

Do you think, after you've tested the properties of each type of plastic, that you'll be able to figure out which is which without the benefit of the recycling code?

# Materials You'll Need for This Project

There are some materials you'll need for this project in addition to the six types of plastic you'll be testing. First, however, be sure that you do have samples of plastics coded with identifying numbers 1 through 6.

The rest of the necessary materials, listed below, can be found in a grocery store if they're not already in your home.

- ◆ Isopropyl rubbing alcohol
- ◆ Light corn syrup
- ◆ Water
- ◆ Metric or standard measuring spoon
- ◆ Four small, plastic cups or bowls
- ◆ Candle and matches for flame test
- ◆ Kitchen tongs
- ◆ Pail of water
- ◆ Fine-tipped permanent marker

To prepare for your experiment, you'll need to cut five samples from each of the plastic containers you've gathered. You need to have five small pieces of type 1 plastic, five of type 2, and so on, through type 6. All together, you'll have 30 little pieces of plastic, divided in six groups of all-same pieces, each one between a one- and two-inch square.

# Conducting Your Experiment

The rubbing alcohol, corn syrup, and water are necessary to create a substance of known density. You will mix them together in different combinations of ingredients, and this will allow you to test the density of each type of plastic.

Follow these steps:

1. Using a fine-tipped permanent marker, write a "1" on each of the five sample pieces taken from the item you collected that was identified as plastic number 1.

2. Write a "2" on the pieces from plastic-type 2, a "3" on those from type 3, and so on, until all the small pieces are marked with the number that corresponds with the type of plastic from which they were cut.

3. Observe the color and clarity of each piece of plastic. Write these observations on the first chart found in the next section, "Keeping Track of Your Experiment."

4. Determine if your sample is soft and pliable, or rigid and hard. Put a check mark in the appropriate column of the same chart.

### Scientific Surprise

Although most plastic items can be recycled, many municipalities collect only certain types of plastic (usually PET and HDPE). That means there is a lot of plastic out there that isn't being recycled.

5. Prepare the solutions listed in the box below, each in its own bowl. These solutions will be used to test the relative density of each of the six different plastics. If the type-1 plastic sinks in all four liquids, for instance, it means that its density is greater than the density of each of the liquids. If it sinks in two of the liquids but floats in the other two, then its density is somewhere between the density of the heavier liquids and the lighter ones. Label each bowl as A, B, C, or D.

| | **Solutions for Testing Density of Plastics** | |
|---|---|---|
| **Solution** | **Mixture of Isopropyl Rubbing Alcohol, Light Corn Syrup, and Water** | **Density in g/ml** |
| A | 5 ml 70% isopropyl alcohol + 2 ml water | 0.91 |
| B | 4 ml 70% isopropyl alcohol + 2 ml water | 0.93 |
| C | water | 1.00 |
| D | 1 ml light corn syrup + 1 ml water | 1.16 |

6. Place one of the five sample pieces of type 1 plastic in the first solution. Record in the second chart found in the next section if the sample sinks or floats.

7. Continue testing the density of the same type of plastic in the other three solutions. When finished, you should have one plastic sample from each group left over.

8. Repeat step 7 for the other samples. You will be using four sample pieces of each different plastic for this density test.

9. Using tongs, hold the last sample from each group over a lighted candle, and observe and record the color of the flame and the color of the smoke in the first chart in the next section. Discard the plastic in a pail of water to ensure that all flames will be extinguished.

10. Repeat step 9 with the remaining plastic sample from each group.

**CAUTION**

**Explosion Ahead**

Be sure to use care when performing this part of the experiment. Work in a well ventilated area of your home, or, better yet, outside. Take care not to breathe in the smoke produced by the burning plastic.

# Keeping Track of Your Experiment

You can use the charts shown below to record your observations from the experiment. Or you can make your own charts, if you prefer.

| Plastic Sample | Color | Clarity (Clear or Opaque) | Soft and Pliable | Hard and Rigid | Flame Color | Smoke Color |
|---|---|---|---|---|---|---|
| 1 | | | | | | |
| 2 | | | | | | |
| 3 | | | | | | |
| 4 | | | | | | |
| 5 | | | | | | |
| 6 | | | | | | |

*Use this chart to record the color and clarity of your plastic samples, whether they're hard or soft, and the colors of the flame and smoke they created when burned.*

| Plastic | Solution A | Solution B | Solution C | Solution D |
|---------|------------|------------|------------|------------|
| #1 PET |  |  |  |  |
| #2 HDPE |  |  |  |  |
| #3 PVC |  |  |  |  |
| #4 LDPE |  |  |  |  |
| #5 PP |  |  |  |  |
| #6 PS |  |  |  |  |

*Use this chart to record the densities of the various plastic samples.*

Once you've recorded all your information and observations, you'll be ready to determine if you're now able to identify different plastics by considering their characteristics and properties.

# Putting It All Together

Think about doing the experiment again, but this time having samples from each group of plastics and not knowing which is which.

Do you think that, knowing what you know from the experiment just completed, you'd be able to identify each sample properly if you again tested the physical and chemical properties of each?

If you're not sure, give it a try!

# Further Investigation

Do a little research and learn how polymers are manufactured. Find out what type of process or chemical reaction is used to build these polymers.

If you're interested in the impact that plastic has on the environment, you could learn more about how each type of plastic, including Styrofoam, is recycled.

# Other Great Earth Science Projects

If you're interested in working with another project or two that deals with earth science, you could consider one or both of the suggestions in the following sections.

The first is a fairly easy but really interesting project that allows you to measure how much oxygen is in the air. In the other you test the acidity of rain and snow.

## How Much Oxygen Is in the Air?

When you step outside and take a big breath of fresh air, do you have any idea what you're breathing in? The atmosphere is made up of a variety of gases, including nitrogen, oxygen, argon, carbon dioxide, neon, helium, methane, and so on. The primary ingredient is nitrogen, followed by oxygen.

You can do your own test of how much oxygen is in the air by making a controlled environment and using a common material to pull the oxygen out of the air. You'll need some of those hand-warmer pouches that outdoorsy types use in the winter. You can find them at sporting goods or hardware stores, and they're generally inexpensive.

> ### Scientific Surprise
>
> Nitrogen is the main ingredient of the atmosphere, logging in at about 78 percent of all the stuff up there. Oxygen comes in second at about 21 percent.

The hand warmer pouches contain various materials that are sealed in plastic. When the materials are shaken together and exposed to the air, which occurs when you open the package, they react with one another and begin to oxidize. The oxidation causes heat. Some popular brands of hand warmers are Hot Rods and Heat Factory.

Fasten a pouch to the inside bottom of a tall, straight glass. You can do this with heavy tape. Then invert the glass onto a shallow plate that contains water.

What will happen is that the materials contained within the hand warmer will deplete the oxygen inside the glass. The other stuff—the nitrogen, carbon dioxide, argon, and so forth—will stay in there, and the oxygen will be replaced with water that will rise from the plate. You'll be able to tell how much oxygen was in the glass by the percentage of water it contains after all the oxygen is used up.

The simple science behind this experiment is that the hand warmers function due to a chemical reaction. The pouches contain iron, which, when exposed to oxygen, oxidizes and makes heat. When the iron oxidizes—or rusts—it devours the oxygen in the air. You can mess around with this idea and perhaps come up with an alternative you like better.

The project suggested here, however, is interesting and fairly dramatic without being very difficult to put together.

## Which Is More Acidic, Snow or Rain?

You've heard of acid rain, but what about acid snow? It makes sense, of course, that if rain becomes acidic as it falls through the atmosphere, the same would occur when it snows.

Do you have an idea, though, if there would be differences between the two? Do you think one might be more acidic than the other? You can measure the acidity of rain and snow in your area using the pH scale and pH paper, which is available through scientific supply companies (see Appendix C) or in some pharmacies.

Acid rain, snow, or sleet is precipitation that is more acidic than pure water, which has a pH of 7.0. Normal rain contains carbon dioxide, which makes it a little more acidic than pure water. The pH of normal rain is about 5.5. True acid rain, however, can have a pH that's much lower. Remember that the lower the pH, the more acid the rain.

There's been a lot of research conducted on acid precipitation, which has been found to cause harm to lakes and the creatures in them, to forests, and to statues and buildings.

## The Least You Need to Know

- ◆ Plastics can be identified by their physical and chemical properties.

- ◆ There are seven types of plastics identified for the purpose of recycling.

- ◆ Different types of plastics have different densities, which can be tested in a solution with a known density.

- ◆ Because the results of your experiment are dependent on your observations, be sure to pay close attention and record all observations accurately.

- ◆ Many people believe that the air they breathe is primarily oxygen, but the atmosphere actually is made up of a variety of gases, of which the primary ingredient is nitrogen.

# Chapter 21

# How Can You Use Physics to Become a Better Hitter (and Other Great Physical Science Projects)?

## In This Chapter

- ◆ The concept of neurological pathways
- ◆ Understanding age and pathway development
- ◆ The physical and mathematical mechanics of baseball
- ◆ How mass and velocity contribute to energy
- ◆ Increasing your strength and speed to play better ball
- ◆ Getting more comfortable with physics

Probably all baseball and softball players—everybody from Little League to Major League—would like to improve their hitting distances. Wouldn't it be great to slam a ball across the outfield every time you stepped up to bat?

So what are you going to do to give the ball more launch the next time you're up to bat?

On June 3, 2003, Sammy Sosa, the great Chicago Cubs hitter, was caught with some cork imbedded in one of his bats. The point of that, of course, is to make the bat lighter so that the player can get it around faster to the ball.

Sosa, who's hit more than 500 home runs during his years in Major League Baseball and is certain to become a member of the Baseball Hall of Fame when he retires, said he accidentally grabbed a bat he used to put on home run displays for fans during batting practice, and mistakenly used it during a game against the Tampa Bay Devil Rays.

Sosa was suspended for seven games, but, the lesson is that cheating is an unacceptable method of improving your runs-batted-in record.

In this chapter, we'll explore the basic physics of baseball, and try to figure out a way for you to hit more home runs for your team—without resorting to stuffing cork into your bat.

# So What Seems to Be the Problem?

The problem that you'll attempt to solve during the course of this science fair project is how you can hit more home runs in your baseball games.

Before you start, though, there are a couple of basic premises that you need to understand.

The first is that, in order to do well at most activities, you need to practice correctly, and spend a lot of time perfecting your technique. This pertains not only to baseball, but all activities, such as ballet, baton twirling, singing, piano playing, basketball, and golf just to name a few.

**Standard Procedure**

Check out a biography of nearly any of your favorite athletes, and you'll read that they started playing the sport at a very young age. Think Tiger Woods, Michael Jordan, or Venus and Serena Williams. Repeatedly, you'll see pictures of them as small children, diligently practicing their sport.

Even really great natural athletes need to train and practice in order to be the best.

The second fact you need to keep in mind is that starting an activity at a very young age gives you a huge advantage over those who wait until they're older to begin.

When you're young, your brain develops neurological pathways that help your body to engage more easily in an activity, such as baseball. After a time, your brain and body know exactly what to do when a ball comes across the plate, or when you need to snag a quick out at first. Your body reacts reflexively, without having to wait for your brain to tell it what to do.

The game becomes almost an automatic routine. Your body doesn't have to think much about hitting the baseball, it just does it.

Most people develop these neurological pathways with everyday activities, like walking. You had to learn how to walk, just like you had to learn how to play the guitar or play tennis. Once you learned, walking became second nature. You probably don't need to think about how to walk—you just do it.

Motor skills, which are learned physical movements, are driven by long-established patterns in the brain. Once you learn to walk, walking becomes an established pattern. Same goes with hitting a baseball.

If you wait until you're 30 to start learning to hit a baseball, or a golf ball, or whatever, you're at a disadvantage. That's because your brain won't accept the pathways that make the game second nature as easily or as well as a younger brain. To be great at baseball, the motor skills should have been set in place a long time ago.

> **Scientific Surprise**
>
> Sorry about this, but if you're just starting a sport at the middle school or high school level, you won't ever have the speed and reflexes of an athlete who's been playing that sport since he or she was five or six. The pathways just don't form as well or as quickly when you're a teenager as they do earlier in life.

If you haven't been playing intense baseball since the time you were four or five years old, however, don't despair. By learning and applying a little physics to your game (along with a little work on your part), you can help achieve a more automatic response when the ball comes across the plate to you.

You'll probably never hit as many home runs as Sammy Sosa, but you can improve your game and increase your chances of knocking the ball over the fence.

If you want to, you can use the name of this chapter, "How Can You Use Physics to Become a Better Hitter?" as the title for your science fair project.

Or you could use one of the titles suggested below.

- ◆ Improving Your Chances of Hitting a Home Run
- ◆ What Do You Need: A Bigger Bat, or a Faster Bat?

# What's the Point?

The point of this project is to help you understand the physical and mathematical mechanics of hitting a home run.

What's necessary in order for you to hit a ball absolutely as far as you can? One thing is a good pitch. The other is energy. A lot of energy.

When you hit a baseball, energy is literally transferred through the bat to the ball. The more energy contained in the bat, the farther the baseball will go.

The official scientific equation for this concept of energy is written below.

Energy = ½ times the mass of the bat times the velocity of the bat (speed in a specific direction) squared

Or more simply:

$$E = 0.5 \ mv^2$$

### Scientific Surprise

Inserting cork into a bat makes the bat only a few ounces lighter, but those few ounces can go a long way in helping a player to get the bat around faster to hit a speeding ball.

### Standard Procedure

Remember that the speed we're referring to at this point isn't the distance the ball travels per second, but the distance you move the bat around as you hit the ball.

So what does this mean to you and your baseball team? The first part of the equation tells you that if you increase the mass of the bat, the bat will potentially have more energy. The heavier the bat is, the farther the ball should travel.

The trouble with that is, if you've had experience with baseball, you know that swinging a heavy bat isn't the easiest task, let alone swinging it quickly. So what do you do?

Fortunately, increasing the mass of the bat is only one way you can increase the energy with which you hit the ball. You also can learn to swing the bat faster. If you check out the equation again, you'll notice that the velocity is squared. That means that doubling the speed at which you swing the bat results in a fourfold increase in energy. In other words, it gives you a lot more bang for your buck.

To better understand this equation, we'll analyze it by varying the weight of the bat and the speed of the swing.

Let's say that your bat weighs 32 ounces, and the speed of your swing is 50 feet per second.

With a 32-ounce (two-pound) bat and a 50-feet-per-second swing, the equation reads like this:

$$E = (0.5)(2 \text{ pounds})(50 \text{ft/sec})^2 = 2,500 \text{ units of energy applied to the ball}$$

Now look at the equation when we've doubled the weight of the bat, while keeping the speed of the swing the same:

E = (0.5) (4 pounds) (50ft/sec)² = 5,000 units of energy applied to ball

Hang in there, now, this is almost over. The last way we'll look at the equation is to go back to the original 2-pound bat, but we'll double the speed of your swing:

E = (0.5) (2 pounds) (100ft/sec)² = 10,000 units of energy applied to ball

The equation clearly shows that doubling the speed of your swing is a much more effective means of improving your chances of hitting a home run than doubling the size of your bat.

Because the velocity is squared, but the mass isn't, you get considerably more energy applied to the ball by increasing the speed of your swing.

The challenge you face by doing this science fair project is to test whether you can increase both your strength and the speed of your swing, resulting in better hitting.

# What Do You Think Will Happen?

If you're able to apply the concepts necessary to hit a home run, then you can hit the ball farther and improve your batting average.

You'll need to practice swinging the bat faster in order to increase the distance you're able to hit the ball. Of course, you'll also need to make contact with the ball. The fastest swing in the world isn't much good unless its energy gets transferred to the ball.

Don't worry, though. As you repeat the movements of swinging faster and hitting the ball, your brain pathways will help these motions to become more reflexive and automatic for you. It won't happen as quickly as it might have when you were five or six, but you'll see a big difference.

### Scientific Surprise

A major league pitcher having a good day can fire a ball over the plate in just half a second. Clearly, that leaves little time for the batter to decide on and execute a swing.

Professional baseball players use their strength and their speed to try to hit the ball consistently, and to hit it far. A stronger player can use a heavier bat *and* still swing faster than a player who isn't as muscular. Those players who aren't as strong need to use a lighter bat and concentrate on swinging it as fast as they can.

### Standard Procedure

Most major league players get a hit only once out of every four times at bat. So don't feel bad if you're not hitting home runs every time you're at the plate.

Practicing batting will help you to increase the speed at which you swing. Working out with weights will increase your strength. Those two things, combined, will make a big difference in the amount of energy you get behind the ball.

For this experiment, you'll practice batting every day, and lift weights a couple of times a week. You'll need access once a week for eight weeks to a batting facility with an automatic pitching machine and a device that measures the speed at which you hit the ball. The speed of your hit will indicate whether or not you're increasing the amount of energy you're transferring from the bat to the ball.

*This experiment should help you improve the speed at which the ball you're hitting leaves the bat.*

### Explosion Ahead

Don't start lifting weights if you don't know what you're doing. Lifting weights that are too heavy or using improper technique can cause injury.

If you can get to such a facility more often than once a week, that's fine. If not, however, an outdoor batting area will suffice on the days you don't have to measure your speed.

You'll also need access to some weights. If possible, work with the trainer at your school or a professional at a gym to help you set up a lifting schedule that's right for you.

Most schools have weight rooms that are open to students at certain times. You'll need to lift weights two or three times a week and practice your swing daily in order to increase your muscle mass and improve your speed.

Keep track of how much you're lifting, and the number of repetitions you do for each exercise.

You'll need someone to go with you to the batting cages once a week to record your speeds. This will allow you to compare the speed of your swing at the beginning of the experiment and the end of the experiment.

The control in this experiment will be how far you hit the ball before you begin practicing and working out. You can gauge that by your score at the batting cages.

The variables are the measures you take to increase your strength, and to, hopefully, increase the speed of your swing. At the end of the experiment, you'll see whether you're able to transfer more energy from the bat to the ball, both by using a bigger bat and by swinging faster.

So go ahead and venture a hypothesis of whether or not you'll be hitting better and farther after eight weeks of practice and lifting. What do you think?

**Standard Procedure**

This is a perfect science fair project to do as a team, if your school allows it. Getting a buddy to do the same experiment will enable you to work together and help each other. Plus, it will make the project a lot more fun.

# Materials You'll Need for This Project

You'll need the following materials for this project:

- A baseball bat

- A baseball

- Access to batting cages that provide an automatic pitching machine and a device that measures the speed of your swing

- Weights, either free weights or a weight machine

- Advice from a trainer or experienced lifter

# Conducting Your Experiment

Follow these steps:

1. Consult with a trainer and begin a weight-lifting routine that's tailored to your physical strength and condition. Be sure the trainer understands your goal so he or she can suggest specific exercises for you.

2. Keep a log of the exercises you'll be doing, the amounts of weight you'll use for each exercise, and the number of repetitions completed for each exercise. You can use the chart found in the next section, "Keeping Track of Your Experiment," if you wish.

3. Practice daily at swinging the bat as fast as you can, while still maintaining control.

4. Once a week, visit the batting cages and record the rate of speed at which you hit the ball.

# Keeping Track of Your Experiment

Record your exercise information on these charts, or make your own charts, if you prefer. Be sure to record your information each time you exercise.

Ideally, have someone accompany you to the batting cages and record your speeds. Have your buddy write down all your speeds for that session so that you can calculate the average speed and write it on the chart.

You will need to make charts similar to these for each different exercise in your program.

Use the following charts to keep track of your average bat swing speeds.

| Week | Weights | Repetitions | | |
|---|---|---|---|---|
| | | Set 1 | Set 2 | Set 3 |
| 1 | | | | |
| 2 | | | | |
| 3 | | | | |
| 4 | | | | |
| 5 | | | | |
| 6 | | | | |
| 7 | | | | |
| 8 | | | | |

*Keep track of the progress you're making on a chart similar to this one.*

**Average Bat Swing Speed Chart**

| Week | Feet/Second | Meters/Second |
|---|---|---|
| 1 | | |
| 2 | | |
| 3 | | |
| 4 | | |
| 5 | | |
| 6 | | |
| 7 | | |
| 8 | | |

*Use this chart, or a similar chart, to record your average bat speed.*

# Putting It All Together

When you've completed the experiment and recorded all the information, think about how much you improved over the eight-week period. Did you see a difference in muscle mass, speed and batting average? Do you sense that hitting the ball has become more reflexive than it was when you first started the experiment? Are you able to use a heavier bat, which also will increase the amount of energy you're transferring from bat to ball?

If you said yes to the questions above, then you proved several things during the experiment. You proved the equation to be correct—that increased mass and velocity result in increased bat speed, which transfers more energy to the ball. And, you proved that neurological pathways—although they don't occur as readily in teens and adults—can contribute to better hitting.

# Further Investigation

If you want to keep working on this idea, you could observe, and possibly interview, really good players at as many levels as possible. Find out how old they were when they began playing, if they follow a workout routine, the weight of the bat they use, their batting averages, and so forth.

Gather as much information as you can, analyze it, and see what patterns you come up with.

# Other Great Physics Projects

Some people get nervous when faced with a physics project, because they perceive physics to be something difficult and complicated.

If you read through the experiment outlined earlier in this chapter, however, hopefully you realized that physics applies to a lot of everyday activities and real-life problems. Very basically, physics can be described as the study of motion, matter, energy, and force.

No doubt you can think of many events and circumstances that involve those four areas.

The rest of this chapter describes two more projects relating to physics that you can do, if you want to learn a little more about this area of science. One experiment compares the abilities of different types of light to penetrate various materials, while the other explores the issues of momentum and friction.

## Can a Visible Light Outshine Infrared Radiation?

In this comparative project, you'll be looking at two types of electromagnetic radiation.

Electromagnetic radiation is energy that travels in transverse waves, and is measured in wavelengths.

A wavelength is measured from the top of one crest to the top of the next crest. The shorter the wavelength is, the more energy it contains.

Gamma rays have the shortest wavelengths and radio waves have the longest wavelengths. Just as the names imply, these waves include magnetic and electrical components.

Electromagnetic radiation includes forms of energy ranging from visible light to radio waves, microwaves, and gamma rays.

These different energies are part of the electromagnetic spectrum, of which only a very small part is the visible spectrum of light containing all the colors in the rainbow.

The infrared radiation that comes from a remote control device, such as the TV clicker, has a larger wavelength than visible light, although we can't see it with the naked eye. On the other side of visible light is ultraviolet radiation from the sun. That's the stuff that causes sunburn and skin cancer.

These wavelengths are shorter, and therefore they have more energy than infrared radiation or visible light. Again though, you can't see that type of radiation with the naked eye.

For this experiment, you'll use an ordinary flashlight, a TV remote control and a few variables such as a glass of water, a glass of milk, sheets of paper, and a piece of plywood that's as large as your TV screen.

To compare these two types of radiation, you'll need to set up a testing area in a room that contains a television.

Place a piece of tape on the floor 8 feet (2.4 meters) away from the TV. This is where you'll stand to conduct each test of this experiment.

Darken the room as much as possible, and make sure the TV is turned off.

You'll conduct the following series of tests separately for the remote control—the source of infrared radiation—and the flashlight—the source of visible light.

Conduct each test with the flashlight first, recording your observations in a journal. Then perform the same test, using the remote control. Again, record your observations.

All tests are performed with you standing at the 8-feet tape mark.

- ◆ Test 1: Shine the flashlight at the TV screen and record what you see. Repeat with the remote control.

- ◆ Test 2: Hold a glass of water directly in front of each device and record your observations when you turn each of them on.

- ◆ Test 3: Hold a glass of milk in front of each, turn each one on, and record your observations.

- ◆ How do the milk and water affect how we see the light? Can the TV still be turned on if the radiation is passing through glass and a liquid?

- ◆ Test 4: Hold a piece of paper in front of each device and turn them on. How many pieces of paper are needed to completely block the visible light from shining onto the TV screen? How many pieces of paper are needed to prevent the infrared radiation from penetrating through the paper so the TV won't turn on?

- ◆ Test 5: Have someone hold the piece of plywood directly in front of the TV screen. Make sure your helper is to the side of the plywood, not in front of it. Turn on each light. Record what you see.

- ◆ Test 6: Continue to decrease the distance between the plywood and the flashlight/remote control in 2-foot increments. Turn on the device and record your observations.

As you should have been able to observe, not all radiation is created equal. Can you think of other ways to test the properties of these two types of radiation? What other variables could you use? How far away can you be from the TV and still turn it on with the remote control? What is its working range? Can you still see the light from the flashlight on the TV screen at that distance?

Create a chart that shows your observations from the tests. What could account for the differences? Do some research before you begin this project to gain a better understanding of electromagnetic radiation.

## How Do Different Surfaces Affect the Momentum of Marbles?

You know what the word *momentum* means, right? There are all sorts of situations in which momentum comes into play, and you encounter them daily. When you whack a ball with a tennis racket, you're transferring momentum from the racket to the ball, moving the ball in a certain direction.

If you run fast down a hill and find it difficult to stop at the bottom, you say you had a lot of momentum behind you. Momentum, simply put, is the mass of a moving body times its velocity, or speed in a specific direction.

If friction is not a factor, momentum will pass from one object to a second and then to a third at a constant rate. The total amount of momentum transferred would remain constant from the first object to the last.

**Basic Elements**

**Momentum** is the mass of a moving body times its velocity. It can be transferred from one object to another.

In the physics experiment described here, you'll apply the concept of momentum using marbles, meter sticks, a stopwatch and different surfaces. You can conduct the experiment on surfaces such as a hardwood floor, linoleum, cement, indoor/outdoor carpeting, plush rug or any other flat, horizontal surface on which a marble can roll.

What you'll do is observe the momentum and calculate the velocity of moving marbles. Take a guess at what you think will occur on various surfaces as momentum is transferred from one marble to another. The formula to calculate velocity is shown below.

$$mv_i \times m''v''_i = mv_f \times m''v''_f$$

m = mass of the first object

$v_i$ = initial velocity of the moving object

m″ = mass of the second object

$v''_i$ = initial velocity of the second object

$v_f$ = final velocity of the second object

$v''_f$ = final velocity of the second object

Begin your observations by taping two meter sticks to a floor at a width just wide enough for one marble. Place two marbles in the middle of the "track" 10 cm apart from each other. Flick the first marble so it hits the second one. How does the velocity of the first marble change? What happened to the second marble?

Next, put two marbles in the track next to each other. Take a third marble and set it about 10 cm from the others. Flick the third marble at the other two. What happened to each marble?

Could you tell that momentum was being transferred? Repeat the procedure on several different surfaces. Does the material from which the surface is made affect the velocity of the marbles? If so, why do you think?

Use the preceding formula, a stopwatch, and a ruler to measure the distance the marbles rolled when momentum was transferred to them. Calculate the velocity of the marbles before and after their collisions on different surfaces, both smooth and on rugs or carpeting. Doing so will give you mathematical data to support your hypothesis.

## The Least You Need to Know

- Patterns set in the brain by repetition can lead to reflexive motor skills.
- These pathways to reflexive motor skills are formed more quickly and easily in young children than in teenagers or adults.
- Baseball is heavily influenced by physical and mathematical factors.
- Mass and velocity are both important to energy, but velocity is the more important of the two.
- Increasing your strength and the speed of your swing should result in improved batting.
- Physics makes many people nervous, but it doesn't need to be overly difficult.

# Part 5

# Projects to Really Wow 'Em

Sometimes you just feel like having a little fun and doing something out of the ordinary. Like making an egg you can bounce, or creating your very own lava lamp.

The projects outlined in Chapters 22 through 27 tend to be a little offbeat and crazy—and lots of fun to do!

That's not to say they shouldn't be taken seriously, as each one provides a valuable opportunity for learning. But, trust us on this, a "rubber" egg is a lot of fun to play with!

Just one thing about that egg … don't play with it in the house!

# How Explosion Proof Is Your Sandwich Bag?

## In This Chapter

- ◆ Identifying the problem you hope to solve
- ◆ Using observations and experience to make an educated guess
- ◆ Gathering materials to use in your experiment
- ◆ Conducting the experiment
- ◆ Recording your results

You've probably had at least one bad experience involving a sandwich bag. These sandwich bag disasters can take many forms, depending on what's in the bag, where the accident occurs, and so forth.

Let's just say that putting a really messy food (think salad with ranch dressing or even a particularly gooey PB&J sandwich) in an unreliable bag is inviting disaster in your lunch bag, and possibly on your lunch table.

Given that distinct possibility, having the right sandwich bag becomes an important issue. Your entire lunchroom status could rest on it!

So how are you going to figure out which sandwich bag will hold up best under tough conditions? In this chapter, you'll learn about a fun experiment that will help you determine which bags are best—and worst.

# So What Seems to Be the Problem?

The practical problem, as just stated, is that a bad sandwich bag can be a hassle at best, and a calamity at worst. The scientific problem, or question that you'll be attempting to answer during the course of this project, is which brand of zipper-lock sandwich bag can withstand the pressures of a chemical explosion.

While a chemical explosion sounds pretty dramatic, you needn't be alarmed. Your experiment will be conducted under controlled conditions, and the worst that will happen is that you'll end up making a bit of a mess. For that reason, it's highly recommended that the experiment for this project be conducted outside.

**CAUTION**

**Explosion Ahead**

In this experiment you'll mix an acid and a base to produce carbon dioxide gas. It's fun to see what happens when you mix materials together, but *never* mix materials without supervision. Mixing certain materials can cause toxic gasses that can harm you or even kill you.

What you're actually going to do is combine vinegar and baking soda, which will cause carbon dioxide gas to form. Carbon dioxide gas forms when an acid (the vinegar) is mixed with a base (the baking soda). The gas will fizz and fill up the bag, perhaps causing it to explode. Greater amounts of carbon dioxide gas will exert more pressure on the sides of the bag, eventually forcing them open.

By the time you finish this project, you'll have identified the type of sandwich bag that holds up best under pressure. And you will have solved your own potential problem for lunchroom embarrassment.

You could use the title of this chapter, "How Explosion Proof is Your Sandwich Bag?" as the heading for your science fair project. You also could use one of these other names as your title:

◆ Which Sandwich Bag Best Withstands a Chemical Explosion?

◆ How Does Your Brand Stand Up to Pressure?

Now that you've got a handle on the problem you're attempting to solve, take a few minutes to consider the point of what you're doing.

# What's the Point?

While it might sound like fun to create a chemical explosion inside a plastic bag, many people might ask, "What's the point?"

The point is that you'll be using the scientific method (as described in Chapter 4) to find out what brand of sandwich bag is toughest. You'll conduct an experiment to discover the answer to a problem, and you can apply the results of the experiment to your own life.

A trip to the grocery store can be a very expensive venture. By the time you've grabbed some bread, some meat, milk, laundry detergent, orange juice, shampoo, spaghetti, ice cream, paper towels, cookies, eggs, salad and vegetables, fruit, mustard, pickles and sandwich bags, you've committed yourself to spending some serious money.

**Basic Elements**

**Store-brand** products are those made especially for the retailer that sells them. They're usually cheaper than national brands (sometimes called **name brands**) because retailers don't pay for advertising and other extra costs. Store brands also are known as retailer brands or private label brands.

The experiment in this project will reveal whether those *name-brand* sandwich bags (the ones you see advertised on TV) are really better than the *store-brand* bags, which tend to cost much less. Maybe you'll find out that the store-brand bags are just as good as the more expensive ones, and you'll be able to save the grocery buyer in your house some money.

You'll be using a store-brand sandwich bag as your control, and name-brand sandwich bags as your variables.

# What Do You Think Will Happen?

Every project that employs the scientific method (as all projects should) requires that you come up with a hypothesis. As you know, a hypothesis is an educated guess concerning the outcome of your experiment.

In this project, you need to guess, or predict, which of the sandwich bags you'll be using will best withstand the pressure of a chemical reaction. You can do this randomly and hope for the best, or you can (as is the correct way) carefully consider what is observable and make a scientifically based prediction.

### Standard Procedure

When trying to come up with a hypothesis, take into consideration any previous experiences (good or bad) you may have had with sandwich bags. Do you already know that one type is better than another because of something that's happened to you? If so, be sure to factor that into your prediction.

Take a few minutes to examine the sandwich bags that you'll be using in your experiment. Do some of them seem to seal better than others? Does the zipper lock on any of them seem to not close properly? Are some of them made of thicker plastic than others? Do the seams on any of the bags appear weak?

By examining the bags and thinking about the apparent qualities of each brand, you should be able to make a reasonable calculation of your results.

# Materials You'll Need for This Project

A great feature of this science project is that all the materials you need are easy to find. In fact, you probably have most or all of them in your home.

You'll need the following items:

- ◆ Three each of five brands of zipper-lock sandwich bags, all the same size (remember that one brand needs to be a store brand)
- ◆ Paper towels or tissues
- ◆ Scissors
- ◆ A ruler
- ◆ A pen or marker
- ◆ A measuring cup
- ◆ A measuring spoon
- ◆ About 4 cups warm water
- ◆ Baking soda (a new box will work best)
- ◆ 7½ cups vinegar (white or apple cider)
- ◆ A stopwatch (preferred), or an easy-to-read watch or clock with a second hand

Be sure to have all your materials gathered and ready to go before you start your experiment. And be certain to keep the various brands of bag separated so that you know which is which. Some bags may have the brand name printed on them, but others won't. If not, you probably should write the brand on the bag before you begin. Your experiment won't be valid if you mix up the bags or lose track of which is which.

# Conducting Your Experiment

While this experiment is fun to do, as noted earlier, it can create a bit of a mess. And while baking soda, water, and vinegar aren't the worst things you could spatter onto a carpet or sofa, it's highly advisable to conduct this experiment outdoors.

Be sure that you leave yourself plenty of time to work through the entire experiment. You'll actually be conducting the experiment three times, which means you'll be testing 15 different bags. That will take a fair amount of time.

Follow these steps:

1. Using the ruler and pen, measure and cut a 12 cm-by-12 cm square out of a paper towel or tissue. You can save some steps later on by layering the towels and cutting several squares at once.

2. In the center of the square, place 1½ tablespoons (about 11 ml) of baking soda.

3. Carefully fold the paper several times to make a square packet, completely enclosing the baking powder.

4. Pour one-half cup (125 ml) of vinegar and one-quarter cup (62 ml) of warm water into one of the sandwich bags.

5. Drop the baking soda packet into the bag, then zip it shut as quickly as you can.

6. Shake the bag for a few seconds. Two or three seconds should be enough.

7. Place the bag carefully on the ground and step back at least 10 feet. You don't need to fear being injured if the bag explodes, but it's best to be out of the way of any material that may escape.

**CAUTION**

**Explosion Ahead**

Be sure to stay away from walls and windows when conducting this experiment. The force of this reaction isn't strong enough to break a window or damage a wall, but dried baking soda can be tough to remove.

**Standard Procedure**

After placing the filled bag on the ground, be sure to back away from it so that you can watch what's happening. If you turn and walk away, you may miss some significant events that you should be observing.

*Observe what's happening with the bag as you back away from it.*

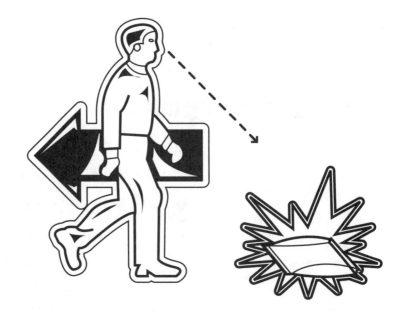

8. Write down the time it takes, in seconds, for the bag to fizz and finally explode. You'll learn what type of chart to use to do this in the next section, "Keeping Track of Your Experiment."

9. Repeat the experiment with each of the other four bags.

10. Repeat the entire experiment two more times.

Doing the experiment three times instead of just once will assure more reliable results.

# Keeping Track of Your Experiment

Perhaps the trickiest part to this science fair project is recording your observations. You'll need to pay very close attention to what's happening to each bag so that you'll be able to mark the exact second that it explodes.

**Standard Procedure**

Using a stopwatch will make it much easier to record how long it takes for each bag to explode, and is highly recommended.

If you find that you're having trouble timing the reaction, you can ask someone to help you. You will need to record the number of seconds that it takes for each bag to explode on a data chart, such as the following one.

This kind of chart is easy to make on your computer or by hand, and will help you to keep your results organized. It also will allow viewers and judges at the science fair to see your results quickly.

| Time in Seconds for Sandwich Bag to Explode | | | | | |
|---|---|---|---|---|---|
| Trial | Bag #1 | Bag #2 | Bag #3 | Bag #4 | Bag #5 |
| 1 | | | | | |
| 2 | | | | | |
| 3 | | | | | |
| Average | | | | | |

*Use this chart to record the amount of time it took for each sandwich bag to explode.*

Once you've recorded all your figures on the chart, you can figure out the average time it took for each brand of bag to explode. You do this by adding the three times for each brand of bag, then dividing that number by 3.

Once you've filled in all the numbers on the chart, you'll be able to analyze your data.

# Putting It All Together

Once you've calculated your results, you can draw a conclusion and find the answer to the problem that you identified when you started the project. You'll know which bag held up the best, and whether the store-brand bag was inferior or superior to the name-brand bags.

# Further Investigation

If you enjoyed doing this project and want to take it a step or two further, you could make some changes to your experiment to see whether or not they would affect its outcome.

Here are some suggestions for variations on the experiment described earlier:

◆ Increase or decrease the water temperature to see if it makes a difference in the amount of gas produced. Remember that the greater the amount of gas generated in the bag, the greater the chance of the bag exploding.

◆ Change the amounts of water, baking soda, and/or vinegar, remembering to keep track of how much of each you use. Using different proportions of the materials could cause different results.

Always feel free to vary an experiment or delve a bit more deeply into a science fair project. Just be sure to keep careful notes of what you've done and what you observe.

## The Least You Need to Know

◆ Knowing which sandwich bag is best will enable you to be an informed shopper.

◆ Vinegar, an acid, and baking soda, a base, are combined in this experiment to form carbon dioxide gas.

◆ This experiment tends to get a little messy, so it's advisable to conduct it outdoors.

◆ You'll need to take care during your experiment to keep the bags organized so they don't get mixed up and make the experiment invalid.

◆ Carefully and neatly recording your results on a data chart will allow them to be viewed quickly and easily.

# 23

# What Size Balloon Rocket Flies the Fastest?

## In This Chapter

- ◆ Learning about the laws of energy
- ◆ Understanding how a balloon rocket works
- ◆ Putting forward your best guess
- ◆ Understanding all the steps before you start
- ◆ Recording results and drawing conclusions
- ◆ Coming up with variations on your experiment

Nothing moves without having some kind of force exerted on it. This is the first of Isaac Newton's three laws of energy: if something isn't moving, it will stay where it is.

The leaves on a tree move frequently. They don't, however, move unless something causes them to. In the case of the leaves, the force that causes them to move is air moving against them. We call this moving air a breeze or wind, depending on its intensity.

The jar of peanut butter in your refrigerator won't move unless force (probably in the form of your hand) is exerted upon it. Unless you or your brother move it, or an earthquake or some other force causes it to tumble from the shelf, it will sit on the shelf indefinitely.

In this chapter, you'll be making a balloon rocket, and testing which size balloon flies the fastest. As with everything else, something must force the rocket to fly. If not, it will stay where it is. In this case, the force will come from the air that's inside of the balloons.

While you can't see the force exerted on the balloons you'll use, it's air. You'll learn exactly how that works a bit later in the chapter. Let's get started.

# So What Seems to Be the Problem?

Balloon rockets are fun and easy to make. If you haven't made one before, you'll be surprised at how simple it is, and how much fun it is to use.

Basically, a balloon rocket uses pressure to move forward. That pressure is the force exerted upon the balloon to propel it across the room.

The pressure is generated when the air trapped within the balloon is released. You see, when you blow up a balloon, the air you put inside of it is pressurized, because it's contained within the balloon. When it's released, look out! It creates the thrust necessary to move the rocket forward.

What happens, is that the particles of air released from the balloon run smack into other particles of air. When the particles from the balloon hit the group of particles outside of the balloon, all the particles are subjected to forces from those in the other group.

The force generated from this meeting of air particles is an example of Isaac Newton's third law of motion, commonly called "action/reaction." It says that every action has an equal and opposite reaction. The action in this case is the air escaping from the balloon. The reaction is the outside air pushing back against the air from the balloon.

You've probably released air from inflated balloons many times, allowing the balloons to fly around the room. With a balloon rocket, the balloon is attached to a straw, through which string is threaded. The string gets attached to two stationary objects, and,

| Scientific Surprise |
| --- |
| A simple balloon rocket works in much the same way as a real rocket. Instead of using inflated balloons to propel them, however, real rockets use huge engines and super-powered fuel. |

when the air is released from the balloon, the straw moves along the string from one stationary object to the other.

Now that you understand how a balloon rocket works, let's identify the problem that you'll be attempting to solve during your experiment. You know that it's the force of the air as it escapes from the balloon that moves the balloon forward. But what kind of balloon will work best?

Will the size of the balloon cause a difference in the speed of your rocket?

# What's the Point?

Quite simply, the point of this experiment is to learn how to make a balloon rocket that flies faster than that of any of your friends. As you've already learned, balloon rockets aren't hard to make. Practically anyone could figure out how to put one together.

By testing different sizes and shapes of balloons, however, you may be able to build the fastest balloon in your neighborhood, or of any of your classmates.

You'll also get hands-on experience using the scientific method, as explained in Chapter 4. And you'll end up better understanding Newton's law of action/reaction.

The control in this experiment is a 10-inch, round balloon. The variables are balloons of other shapes and sizes. Try to find a good variety so you'll be able to conduct a sound experiment.

**Standard Procedure**

The air trapped inside a balloon will escape whenever it has an opportunity, such as when the balloon closing becomes untied or unclipped, or through a hole. This is because the air inside the balloon has more pressure than the air outside, and is trying to equalize that pressure.

# What Do You Think Will Happen?

All of the balloons you'll use for this experiment will contain air. And, you'll be releasing the air from each of the balloons, causing the same reaction to occur with the air particles outside of the balloon.

So what's your guess about how balloon size will affect this reaction? Do you think that a bigger balloon will cause the balloon rocket to fly faster than a small one?

**Basic Elements**

Rocket engines are different from the engines used in cars, tractors, fans, and so forth. While most engines are used to produce rotational energy that powers the machine, rocket engines are reaction engines that propel using the forces of action and reaction.

It will contain more air, that's for sure. Or, perhaps a long, slender balloon that's more the shape of a real rocket would work best.

If you've never made a balloon rocket before, you'll have to make your best guess (a hypothesis) about which balloon will cause the rocket to fly fastest. Try to consider all the known factors, and use your best common sense. And remember, an incorrect hypothesis doesn't affect your experiment or the outcome of your project. It just means that you'll have learned something different from what you believed was true.

# Materials You'll Need for This Project

All of the materials you'll need for this project are readily available and inexpensive. You should be able to find everything you need in your local drug store, toy store, or discount store.

The materials you'll need are listed here.

- Three or more balloons of various sizes and shapes (remember that one has to be a 10-inch, round balloon; try to find other balloons that are as differently shaped from that one as possible)

- One roll of kite string

- A box of plastic straws (get the straight straws, not the kind with the bendable elbows)

- Tape

- A clothespin or twist tie (such as one used to close a trash bag)

- A stopwatch

You'll need a bit of space to conduct this experiment, so be sure to consider that before you get started. If you can find an open, outdoor area for the experiment, that's great. You'll need to make sure, however, that wind is not a factor, and that there are no trees, shrubs, or other objects to impede the travel of the balloon rocket.

A basement or other open, indoor area would work as well. The length of the kite string you'll use depends on the amount of room you have for the balloon rocket to travel along the string.

# Conducting Your Experiment

Before you begin your experiment, during which you'll measure how long it takes for your balloon rocket to travel along a length of string you've attached to two stationary objects, take a few minutes to read over all the steps listed below.

If you've ever put together a model, or made a cake, you've probably learned that it works best to read through all the instructions before you start actually building or mixing ingredients.

You'll need to either cut a piece of the kite string, or unravel some of it from its spool. The length of the string must be sufficient to run between two stationary objects that are positioned fairly far apart. The amount of string you'll need will depend on how far apart the stationary objects are located. Don't forget to measure the string once you've got it extended between the two chairs or whatever you're using to secure it. You'll need to know the length in order to calculate your results.

**CAUTION**

### Explosion Ahead

Even if you're conducting your experiment inside, be sure there is no moving air from a fan or air vent that could affect the speed that the balloon moves. If there is, it could alter the balloon's speed and make your experiment invalid.

Remember to work carefully and to keep your materials organized.

Follow these steps:

1. Tie one end of the kite string tightly to a secure object, such as the leg of a chair, table, or stool. If you haven't cut the string, tie the end closest to the spool on which the string is contained.

2. Pull the loose end of the string through the straw.

3. Pull the string tightly through the straw, then secure the loose end to another secure object so that the string is at a slant.

4. Blow up the balloon, but don't tie it closed.

5. If you have a clothespin that pinches tightly enough to prevent the air from escaping, you can use it to secure the opening of the balloon. If you don't have a clothespin, or it isn't tight enough, secure the top of the balloon with a twist tie so that the air doesn't escape.

6. Measure the length of the balloon in centimeters. Record the measurement on the first chart in the next section, "Keeping Track of Your Experiment."

*Set up a track similar to this for your balloon rocket to travel along.*

*Measure both the length and the circumference of each balloon.*

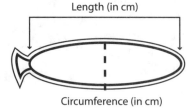

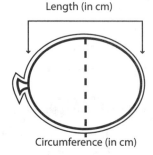

7. Measure the circumference of each balloon in centimeters and record the measurement on the same chart.

8. Tape the top of the balloon to the underside of the straw so that the balloon hangs underneath the straw.

9. Slide the straw with the attached balloon to the lower end of the string, so that the tip of the straw is resting against the stationary object. (Make sure the straw tip touches the stationary object with every balloon you launch—you don't want to give any balloon a head start.) You will be launching the balloon upward.

10. You're ready to launch!

11. Hold the straw and balloon steady at the lower end of the string and release.

12. Start the stopwatch *as soon* as you release the balloon, and stop the watch *as soon* as the balloon reaches the other end of the string. Record this time on the second chart shown in the next section.

13. Repeat steps 1 through 12 using different sizes and shapes of balloons. If you use more than three balloons, you'll need to make another chart, similar to the second one shown in the next section.

For more accurate results, test each balloon three times. Then you can calculate an average of the time it took for each balloon to travel from one end of the string to the other. Record the average times onto the chart.

**Standard Procedure** _____

This experiment will be much easier if you have a friend or family member help you, especially with the timing. Have your friend, sister, or brother stand at the stopping point and yell at the exact moment the balloon arrives there.

**Standard Procedure** _____

Remember that to find the average time it took for each balloon to travel from one end of the string to the other, you add together the times, and then divide them by three. The number you get when you divide the times is the average time.

# Keeping Track of Your Experiment

Once you know the distance the balloon traveled and the average time it took for it to get from one end of the string to the other, you can calculate the speed of your rocket balloon. Use the numbers you've recorded on the following charts to get your findings.

|  | **Length in Centimeters** | **Circumference in Centimeters** |
|---|---|---|
| Balloon #1 |  |  |
| Balloon #2 |  |  |
| Balloon #3 |  |  |

*Use this chart to record various information about the balloons.*

| Time in Seconds for Balloon to Reach End of String | | | |
| --- | --- | --- | --- |
| Trial | Balloon #1 | Balloon #2 | Balloon #3 |
| 1 | | | |
| 2 | | | |
| 3 | | | |

*Use this chart to record the time it took for each balloon to reach the end of the string.*

Taking an average: Balloon # 1

Trial 1: time _____

Trial 2: time _____

Trial 3: time _____

+ _____

Total time _____ divided by 3 = average time

*Use this chart to calculate the average for each balloon.*

# Putting It All Together

Once you've analyzed your data, you'll be able to summarize what you've learned, and you'll see whether your hypothesis was correct.

You'll also be able to figure out which size of balloon caused the balloon rocket to move the most quickly. Did you find that the round balloon flew up the string the fastest? Perhaps it was a long balloon that won the race.

Regardless of your results, you will have reached a conclusion, which is the last step of the scientific method. You will have answered the question posed in your problem, and know if you were right when you formulated your hypothesis.

# Further Investigation

If you enjoyed doing this experiment and would like to take it a step or two further, here are some suggestions.

Tie the kite string to two stationary objects that are located 20 to 30 feet (6 to 9 meters) apart. This will reverse the procedure and allow you to measure the distance each type of balloon travels within a certain amount of time, instead of measuring the time.

With a friend or family member helping, see how far along the string the balloon rocket can travel in 5 seconds. One of you will be in charge of the stopwatch, while the other is responsible for marking the point on the string the balloon had reached after 5 seconds.

Once you've done that, you can calculate the speed of your balloon rocket using the numbers you and your friend have recorded. Remember, speed equals length divided by time. Or you can simply measure the distance the balloon travels until it runs out of air. Watch out for low-flying birds!

Another variation on the original experiment would be to weight the balloon and see if a heavier balloon flies along the string faster than a light one. You can add weight to the outside of the balloon by taping a few pennies onto it.

If you conduct this version of the experiment, use the same size balloons. Your control will be a nonweighted balloon. The variables are different amounts of pennies taped to other balloons.

Feel free to think of other ways in which you can vary the balloon rocket experiment.

## The Least You Need to Know

- A balloon rocket is propelled forward when particles of air from inside the balloon react with particles of air outside of the balloon.

- Understanding how a balloon rocket works will help you understand Isaac Newton's third law of energy, which addresses action and reaction.

- It's okay to guess at the results of your experiment, as long as you consider what you know and use common sense.

- The materials needed for this experiment are common, easy to find, and inexpensive.

◆ This experiment involves many steps, so be sure you understand it completely before you start.

◆ There are many ways in which you can vary the experiment.

# Chapter 24

# Which Brand of Rocket Fuel Propels Your Canister Rocket Best?

## In This Chapter

◆ Making sense of rocket science

◆ Understanding chemical reactions

◆ Name-brand vs. store brand

◆ Using caution when caution is due

◆ Using care to track results

Sometimes the best uses for objects are way different from the intended purpose of the object.

Take those little plastic canisters your camera film comes in, for instance. Sure, they're great for holding film, but think about what else you can do with them. You can use them to store small items, like pins or paper clips. You can use them to build things.

You can even use them—read carefully—to make canister rockets. That's right, canister rockets. You probably never realized the potential for such fun with a simple little film container.

So how the heck are you going to make a rocket out of a film container, you ask? You'll read exactly how a little later in this chapter. For now, just trust us when we tell you it's going to be a blast!

# So What Seems to Be the Problem?

The problem you'll attempt to solve during the course of this project is a simple one. Which "fuel" will send your canister rocket highest into the air?

| Scientific Surprise |
| --- |
| Sir Isaac Newton, a British scientist and mathematician who lived between 1642 and 1727, was meant by his family to be a farmer, taking over the work after his father died. Instead, he became one of the most revered scientists of all time. |

This experiment relies on Isaac Newton's third law of motion, which says "for every action in nature, there is an equal and opposite reaction."

Actual rockets are forced upward by burning fuel that produces large amounts of gases expanded by heat. Your film canister rocket, on the other hand, will be forced upward by antacid tablets commonly used to relieve heartburn and other forms of gastric distress.

Your task, if you're up to it, is to determine which brand of antacid tablets best propels your canister rocket.

If you want to, you can use a slightly modified version of the title of this chapter, "Which Brand of Rocket Fuel Propels Your Canister Rocket Best?" Just change the word *your* to *a*.

Or, you could use one of these titles for your project.

◆ Does Alka-Seltzer Fly Higher Than Store-Brand Antacids?

◆ How High Can Your Rocket Fly?

# What's the Point?

Some people would consider it a huge waste of time to determine which brand of antacid tablet will best propel a rocket made from a film canister.

In the name of science, however, we respectfully disagree. The science of rockets has been understood for years, allowing the business of rocketry to reach previously unimaginable heights (no pun intended).

This experiment allows you to experience firsthand the science of rockets, giving you not only a better understanding of rocket science, but of Newton's third law—one of the most important laws of science ever.

And if you're lucky, you'll learn something about the potency of antacid tablets, which could come in handy the next time you overdo it on chili dogs or something.

# What Do You Think Will Happen?

In this experiment, you'll test different brands of antacid tablets, using the chemical reaction that occurs when you drop the tablet into water to launch your canister rocket into the air.

The gas created during the *chemical reaction* will force the canister upward, serving as rocket fuel.

One of the products of the reaction that occurs when you drop an antacid tablet in water is carbon dioxide gas. This is the same gas you're breathing out of your lungs right now.

**Basic Elements**

A **chemical reaction** is a chemical change that produces new and different substances.

When the gas is produced in the container and not allowed to escape, pressure builds up inside the container, as you might imagine. As more gas is created from the continuing reaction inside the container, the pressure exerted on the top, bottom and sides of the container becomes greater and greater.

Eventually, the container can no longer handle the extra pressure being exerted on it from inside, and, you guessed it, the container will burst open.

The first thing to go is the lid of the canister, because it's attached to the container more loosely than the sides and bottom. The lid will be ejected from the body of the canister, and the canister will be launched into the air.

This forward thrust of the rocket is due to Newton's action and reaction principle. The gas being blown out of the open end of the container is the action. The gas hits the ground (in this case the bottom of the canister), and has nowhere else to go. So it pushes back up into the container, forcing it to fly upward. The flight upward is the reaction to the action of the gas.

To demonstrate this theory, you can use any type of antacid tablet in water. For this experiment, however, you'll use different brands of antacid tablets to see if there's a difference in how well they launch your canister rocket.

Maybe name-brand tablets, such as Alka-Seltzer, are more powerful than the store-brand variety, which normally costs less than name brands.

You'll need to find four or five different kinds of antacid tablets, so you may need to visit more than one place if you need to obtain store brands.

### Standard Procedure

Varying the number of antacid tablets and amount of water used will vary the distance the rocket travels. For purposes of this experiment, though, you'll only be comparing brands of antacids, not varying amounts of water.

The control you'll use is the name-brand antacid tablet that nearly everyone knows about—Alka-Seltzer. The variables will be the store-brand or other name-brand antacid tablets.

Needless to say, this is an outdoor experiment. Do not, under any circumstances, try to perform the experiment inside. You'll need to test each brand of antacid tablet three times, after which you'll average the results.

Obviously, you won't be able to get an exact measurement on the height of each launch. To best estimate the height your rocket travels, observe how high into the air it flies by comparing the height to that of a stationary object for which you know the height.

Maybe you have an 8-foot-high basketball net outside your house, or a 16-foot-high flagpole.

So go ahead and make your best guess about the outcome of the experiment. Do you think that the Alka-Seltzer will outperform the store-brand antacid tablets? Or do you think one of the less expensive brands will end up creating the greatest chemical reaction?

### Standard Procedure

If you can only get one or two canisters, that's okay. But, you'll need to rinse them out and dry them thoroughly between each use. Any water left inside could alter the outcome of the experiment.

# Materials You'll Need for This Project

For this experiment, you need the following:

- At least one film canister and its lid (preferably more than one)

- Teaspoon

- Small container of water

- Four or five different brands antacid tablets

Once you've got everything you need, head outside and let the fun begin!

# Conducting Your Experiment

Because you'll be testing four or five different types of antacid tablets, it's important that you be organized about recording your results and keeping track of which brands you've tested.

Keep the data charts that are in the next section, "Keeping Track of Your Experiment," handy while you conduct the experiment.

Follow these steps:

1. Remove the lid from the film canister(s)

2. Place one brand of antacid tablet inside the canister.

3. Add 1 teaspoon (5 ml) water to the canister.

4. Working quickly, put the lid tightly over the canister. Be sure that it's on properly.

5. Set the canister, lid-end down, on the ground.

*Set the canister, lid-end down, on the ground.*

6. Immediately step back at least 6 feet (about 2 meters) from the canister.

7. Wait for about 10 seconds. You will hear a popping noise, and the film canister will launch into the air.

*Pressure within the canister forces off the top and causes the canister to launch into the air.*

### CAUTION
### Explosion Ahead

It is extremely important that you back away from the canister immediately after setting it upside down on the ground at its launching site. It's possible that the canister could take a wrong direction and head toward you rather than go straight upward.

8. Observe how high the rocket travels, comparing the height to a stationary object you know the height of.

9. Estimate the height the rocket traveled, and record the results. You may use a data chart similar to the one in "Keeping Track of Your Experiment."

10. Measure the distance the rocket fell from the launching site, and record it.

11. Thoroughly rinse out and dry the canister if you're going to reuse it.

12. Repeat steps 1 through 11 two more times, so that you've completed a total of three trials for that particular brand of antacid.

13. Repeat steps 1 through 11 with each different brand of antacid tablet.

If a filled canister does not launch after 10 seconds, wait one minute before approaching and examining the container. Usually, the problem is simply that the lid wasn't on tightly enough. If this is the case, start again by repeating the steps of the experiment.

# Keeping Track of Your Experiment

You can use the first chart below to keep track of the estimated height of your canister rockets, and the distance traveled from the launching site. Use the second chart to record your averages.

You can make your own charts, if you prefer.

| Brand of Antacid | Estimated Vertical Height of Rocket | | | | | | Horizontal Distance from Launch Site to Landing Site | | | | | |
|---|---|---|---|---|---|---|---|---|---|---|---|---|
| | Trial | | | | | | Trial | | | | | |
| | 1 | | 2 | | 3 | | 1 | | 2 | | 3 | |
| | ft | m | ft | m | ft | m | ft | m | ft | m | ft | m |
| | | | | | | | | | | | | |
| | | | | | | | | | | | | |
| | | | | | | | | | | | | |

*Use this chart to record the estimated height and the distance your canisters traveled.*

| Brand of Antacid | Vertical Distance | | Horizontal Distance | |
|---|---|---|---|---|
| | Feet | Meters | Feet | Meters |
| | | | | |
| | | | | |
| | | | | |

*Use this chart to record the average height and distance the canister rockets traveled.*

# Putting It All Together

By analyzing the distance the canister rocket travels in the air, and the distance from its launching site to its landing site, you can determine which rocket fuel works the best.

We're not sure if the blasting power of the different antacid tablets corresponds to its effectiveness against heartburn, but it's an interesting question.

# Further Investigation

If you had fun with this experiment and would like to take it a step or two further, you could vary the amount of water you put inside the canister, using the same brand of antacid.

If Alka-Seltzer propelled your rocket highest and farthest, for instance, vary the amount of water you use from 1 teaspoon to ¼ teaspoon, ½ teaspoon, 1½ teaspoons, and so forth.

Remember to conduct three trials for each amount of water you test.

You can also experiment a bit further by adding a few aerodynamic accessories to the canister. Think about making some paper fins, or a nose cone, and taping them onto the film canister.

At the very least, you'll learn something about aerodynamics, and have a good time in the process.

## The Least You Need to Know

◆ Even the most potentially complex theories can be made simple when applied to something you can relate to.

◆ A chemical reaction is a chemical change that creates a new substance, such as a solid and liquid combining to create a gas.

◆ Testing different brands of antacid tablets in this experiment may result in additional knowledge about the brands.

◆ This experiment is only intended to be done outdoors; don't attempt to conduct it inside.

◆ You'll need to watch the flights of the rockets carefully in order to estimate the height achieved by each canister.

# Chapter 25

# What Liquid Mixtures Make the Best Lava Lamp?

## In This Chapter

- ◆ A brief history of the lava lamp
- ◆ How a lava lamp works
- ◆ Materials you'll need to build a lava lamp
- ◆ Experimenting with your experiment
- ◆ Remembering to use caution at all times
- ◆ Taking the lava lamp idea a step or two further

The motion lamp that has become commonly known as the lava lamp was invented and first produced in England as the Astro lamp. It has been a source of fascination to millions of people over the past four decades.

Introduced in 1963, the lava lamp became a symbol of 1960s culture. Its popularity fluctuated in the decades since the 1960s, but the lamp has staged something of a comeback in recent years. Haggerty Enterprises, Inc., the Chicago-based company that manufactures LAVA brand motion lamps, reports that it turns out 400,000 lamps a year to keep up with demand.

Invented by an engineer named Edward Craven Walker, the Astro lamp was at first considered grotesque and offensive by British shopkeepers, who wanted no part of selling the lights in their store. After the 1960s culture caught hold in Great Britain, however, the lamps became hot items.

The motion lamp made its way to America after Adolph Wertheimer and Hy Spector, two entrepreneurs from Chicago, saw it displayed at a German trade show in 1965. They were enthralled with the novelty light, and bought the rights to manufacture it in North America. They renamed it the LAVA brand motion lamp, and the rest, as they say, is history.

Now that you know a little bit about how the lava lamp came to be, maybe you'll be interested in learning to make your own. It won't look exactly like the ones you buy at the mall, but you'll have a good time with the project.

# So What Seems to Be the Problem?

If you're going to make a lava lamp, you'll need to figure out what liquids work best as the "lava." To do that, you first have to understand how a lava lamp works.

*The first lava lamp, which looked much like the one shown here, was introduced in 1963.*

The lava lamp, like the one illustrated here, is heated from the bottom. When the heat from the bottom of the lamp warms the material inside, the denser liquid at the bottom expands and is able to rise to the top of the lamp and displace the liquid there. As the liquids move around inside the lamp, they create the "lava" look.

The trick of creating a lava lamp is to find two or more liquids that are *immiscible—* meaning that they won't mix. The liquids also must be of different densities, or more precisely, different *specific gravities.*

It is these properties that cause the liquids to move in different ways within the container without mixing when they are heated. Edward Craven Walker's original invention contained a mixture of wax, oil, and other substances. (It was a secret formula.)

You, though, will experiment with mixtures of mineral oil with varying amounts of isopropyl alcohol. The problem you'll attempt to solve is how much of each liquid makes the best lava lamp. When you've finished, you should be the proud owner of a functional lava lamp, with which you'll be able to impress all your friends.

You could use the title of this chapter, "What Liquid Mixtures Make the Best Lava Lamp?" as the title for your science fair project. If you want something a little less formal, consider these titles:

### Basic Elements

The word **immiscible** means incapable of mixing or blending, such as in the case of oil and water. **Specific gravity** is the ratio of the mass of a solid or liquid to the mass of an equal volume of distilled water at 4 degrees Celsius.

### Scientific Surprise

The original and best-selling Lava Lamp throughout the 1960s and 1970s was the Century model, which you can still buy today. It's got a gold base with tiny holes, meant to look like stars when the lamp is turned on. The 52-ounce globe contains yellow or blue liquid, and the lava is colored red or white. Very cool.

◆ Mixing Common Liquids to Create a Lava Lamp

◆ The Best Blends for Lava Lamps

Now that you've identified the problem you'll be attempting to solve, you can move on and consider the purpose of this science fair project.

# What's the Point?

Invariably, the question comes to mind of why anyone would even think of spending valuable hours figuring out which combinations of liquids work best in a lava lamp.

Surely there are better ways to spend your time. You could, for instance, clean out your sock drawer or arrange pencils and pens by length and color in your pencil holder!

**CAUTION**

### Explosion Ahead

It is necessary to generate heat during this experiment so that the liquids move about the container. Be sure to use extreme caution while conducting the experiment, and ask for help if you need it. This project is not recommended for younger children, and is probably most suited to teenagers.

So why are you considering this project? Probably because you expect it will be fun and interesting. If that's the case, you're right! It will be fun, and a great learning experience, as well.

Using the control materials of specific amounts of mineral oil and isopropyl alcohol, you'll discover the optimum amount of alcohol that must be placed in the mineral oil in order for the oil to move around within the heated container, creating the lava look.

The variable is the quantity of 70 percent isopropyl alcohol that affects the movement of the oil. In most experiments, your variable would be a different material or altered conditions.

If you were trying to figure out which metal corrodes the fastest, for instance, as in Chapter 19, your variables would be the different liquids in which you immersed pieces of metal wire. If you were trying to figure out which foods molds grow best on, as in Chapter 7, your variables would be different conditions into which the foods were placed.

In this experiment, however, the control and variable are the same materials, only in varying amounts. You're trying to determine what amount of 70 percent isopropyl alcohol works best.

Through your experiment, you'll be able to figure out which combinations of those materials will work best in your lamp. You'll use the scientific method (as described in Chapter 4) to work through the project and come up with an answer to the problem.

So there you are. The point of this science fair project is to get you doing the work of an amateur scientist. Isn't that more fun than cleaning your sock drawer?

# What Do You Think Will Happen?

Before you venture to make a hypothesis concerning the results of your experiment, take a few minutes to consider exactly what you'll be doing and to think about the materials you'll be using.

Take a look at the mineral oil and the two types of isopropyl alcohol (70 percent and 90 percent) you'll be using. Do you notice anything different about the liquids? Does one appear to be thinner or thicker than the others? Are the colors different? Are there any visible variables between the two types of alcohol?

The more you can observe before making your hypothesis, the more educated a guess you'll be able to make. Think about what will happen during the course of your experiment. You'll be using heat to force substances to rise from the bottom of the jar. Which substance do you think may contain the properties to best do that?

After you've spent a little time thinking about the procedures and materials you'll be using, you can make your best guess of the results you'll get.

**Standard Procedure**

This science fair project will be more time-consuming and difficult than most of the others in this book. You may have to do the experiment several times before you find a combination of liquids that makes a successful lava look. Be patient, and have fun.

Just remember that if your hypothesis turns out to be incorrect, you can still have a valid and valuable science fair project. It's more important to answer the question using the scientific method than to correctly guess what your results will be.

# Materials You'll Need for This Project

This project requires some materials that you may not have around your house, and you may have to spend a bit of money to get what you need. Among the materials you need are two concentrations of isopropyl alcohol. The more common kind is the 70 percent concentration, which can be purchased in almost any grocery or drug store and is frequently called rubbing alcohol.

The 90 percent concentration is less commonly seen, and you may have to ask someone at your local pharmacy about getting you some. Some stores keep less requested items behind the counter, so be sure to ask if you don't see it. The other materials you'll need include the following:

♦ 20-ounce (600 ml) glass bottle, clean, with the label removed. (Note: Tall and fairly narrow bottles work better than short, wide ones. A large iced tea or juice bottle of this size is a good choice.)

♦ Permanent marker

♦ Funnel

♦ Heavy-duty scissors or shears

♦ Coffee can (remove plastic lid)

♦ Food coloring

- Lamp base (you can buy this at a home supply or hardware store, probably for less than $5. If you can find one that includes a dimmer switch, you could further experiment by varying the level of heat)

- 40-watt light bulb (no higher)

- Measuring cup (preferably metric)

- Small bottle of mineral oil (you probably can find this in a grocery or drug store)

- 90 percent isopropyl alcohol (you'll need about 6 ounces [150 ml] of this. If you can't find this in your local drug store, ask your druggist if you can order some)

- 70 percent isopropyl alcohol (you'll need about 18 ounces [450 ml] for one experiment. Buy more if you think you might want to repeat the experiment)

Once you have your materials gathered and organized, you'll be ready to begin the experiment.

# Conducting Your Experiment

This experiment contains quite a few steps, and may seem a little confusing at first. Take a few minutes to read over this entire section, and try to visualize what you'll be doing as you read. Make sure as you read that you have all the necessary materials.

Follow these steps:

1. Using the permanent marker, trace the bottom of the bottle onto the center of the bottom of the coffee can.

2. Use the shears to cut out a circle a little bit larger than what you traced.

3. Near the top of the can, cut a small circle through which you can insert a lamp cord.

4. Place the 40-watt light bulb into the lamp base.

5. Turn the coffee can upside down and center it over the lamp base and the light bulb.

6. Feed the lamp cord through the small hole you made in the coffee can.

7. Holding the glass bottle upright, carefully center it on top of the inverted coffee can, over the opening for the light bulb.

8. Using a funnel if necessary, slowly pour 2 ounces (60 ml) of mineral oil into the bottle.

9. Using a funnel if necessary, slowly pour 5 ounces (150 ml) of 90 percent iso-propyl alcohol into the bottle.

10. Let the liquids settle and record your observations. (You can use a chart similar to the one shown in the next section, "Keeping Track of Your Experiment.")

11. Add a few drops of food coloring to the mixture.

12. Turn on the lamp. Be careful, the glass bottle will get very hot.

13. Add 2 ounces (60 ml) of the 70 percent isopropyl alcohol.

14. Let the liquids settle and record your observations.

15. Repeat step 13 until the mineral oil turns whitish in color and becomes suspended in the mixture, causing it to look like lava. Be sure to add only 2 ounces at a time.

16. Calculate the total amount (in milliliters) of 70 percent alcohol needed to achieve this effect.

> **CAUTION**
>
> **Explosion Ahead**
>
> It's very important to use caution while proceeding with this experiment. Many of the components, including the liquid inside the jar, will be very hot, and can cause burns if not handled extremely carefully. Never leave the lava lamp on when you leave the room or are not attending to it.

Making this lava lamp is not an exact science. We can't tell you that adding $4\frac{3}{4}$ ounces of 70 percent isopropyl alcohol will create the exact lava effect that you're looking for.

Although it's recommended in the experiment that you add 70 percent isopropyl alcohol in 2-ounce increments, if you sense the liquid in your lamp is getting close to the point where it will begin to flow in a lava-type fashion, decrease the amount of alcohol you continue to add to just a few drops at a time. Two ounces could turn out to be too much, meaning you'd have to start the experiment again.

# Keeping Track of Your Experiment

This is a fairly complicated experiment, with many steps. It's very important that you keep track of your observations as they occur. You need to know how much of each liquid you used to achieve the effect you were looking for. Don't expect that you'll remember the exact amounts of what you've used if you don't write them down immediately.

Remember that you'll be using both qualitative and quantitative measurements in this experiment. Your qualitative measurements will be those that you observe, but cannot directly measure with tools, such as a ruler or thermometer. Quantitative measurements are those for which you can use tools to get a definite reading in a specific unit of measurement, such as a gram or milliliter.

It's particularly important in this experiment that you accurately record your observations from Step 13. Since the variable is the different amounts of 70 percent isopropyl alcohol, you need to note exactly what amount of that liquid is necessary to achieve the lava look.

You can use the chart below, or make a similar one, to record your observations.

---

**Observations of Adding Isopropyl Alcohol to 60 ml of Mineral Oil**

150 ml of 90 percent isopropyl alcohol

60 ml of 70 percent isopropyl alcohol

60 ml of 70 percent isopropyl alcohol

60 ml of 70 percent isopropyl alcohol

---

Continue the chart if necessary to record your observations for each time you added an additional 2 ounces (60 ml) of 70 percent isopropyl alcohol. Once you've finished the chart, you can calculate the total amount of 70 percent isopropyl alcohol that was necessary to achieve the lava look.

---

**Results of Adding Isopropyl Alcohol to 60 ml of Mineral Oil**

_____ ml of mineral oil

_____ ml of 90 percent isopropyl alcohol

_____ ml of 70 percent isopropyl alcohol

---

If you really want to impress the judges, you could record the final amounts of materials you used on a bar graph (see Chapter 6). This would clearly illustrate the ratios.

# Putting It All Together

During the course of your experiment, you'll discover the ratio of the three different liquids needed to create the lava effect in your lamp.

You will have discovered which mixtures of the liquids used worked best to cause the mineral oil to float through the bottle, and you'll be able to summarize what you've learned, and conclude whether your hypothesis was correct.

And, hopefully, you will have had a great time creating your very own lava lamp. It may not look quite as cool as the ones you can buy, but there's a lot to be said for building something on your own.

**Standard Procedure**

If you want to learn more about lava lamps in general and specifically about how to make them, check out this great website, Oozing Goo. There's even a discussion group where you can exchange ideas with others who have made lamps. It's on the web at www.oozinggoo.com.

# Further Investigation

If you enjoyed building your own lava lamp, you might want to take this project a step or two further. There are several ways in which you might do this.

Using the same lamp model, try mixing other liquids with different densities. Some that you could try include vegetable oils, water, salt water, honey, glycerin, liquid dish soap, liquid hand soap, and castor oil.

If you got a lamp base that included a dimmer switch, you also could vary the experiment by changing the heat level. By doing so, you could determine whether the lamp worked best with the 40-watt bulb at its brightest level, or at lower levels as controlled by the dimmer.

You also could experiment by using bottles or jars of varying shapes and sizes. Perhaps a larger jar would allow the liquid to flow more freely and result in a different look.

There is a lot of online information about making lava lamps, some of which recommends using potentially hazardous materials such as turpentine or antifreeze. We highly recommend that you do not experiment with any substances that could be harmful, and that you work with a partner, when possible.

## The Least You Need to Know

- Introduced in 1963, lava lamps became a symbol of '60s culture, and remain somewhat popular today.

- The "lava" look is created when heat from the bottom of the lamp warms the material inside, causing the denser liquid at the bottom to expand and rise to the top of the lamp, displacing other liquid there.

- You can build your own lava lamp using simple materials such as a coffee can, glass bottle, lamp base, light bulb, and common liquids.

- You may have to experiment a bit during the course of your experiment by adding more alcohol to the mineral oil, a little bit at a time.

- There are many ways to make a lava lamp, but you must be extremely careful because some of them use potentially harmful substances.

# Chapter 26

# How Much Energy Do Different Types of Nuts Contain?

## In This Chapter

- ◆ Understanding how stored energy is converted to heat
- ◆ Using your knowledge to formulate a hypothesis
- ◆ Materials you'll need for your experiment
- ◆ Knowing how to set up the equipment
- ◆ Heating water with burning nuts
- ◆ Other ideas for testing energy

You're probably most familiar with energy in the forms of heat, light, sound, electricity, and nuclear energy. All these forms of energy cause motion or changes in matter. Remember, though, that energy cannot be created or destroyed, only changed from one form to another, according to the law of conservation of energy.

Different units are used to measure different forms of energy. In science, the most common unit used to measure heat is called a Joule (for some reason, heat emitted from an electric fireplace is not measured in Joules, but in British Thermal Units—go figure). Light and sound are measured by the lengths of the waves they generate in their surroundings. Amp (ampere) is the unit for measuring electrical current; radioactivity is measured in curies.

The energy you'll be working with during the experiment in this chapter is stored energy in food, which is measured in calories. You'll actually perform an experiment demonstrating how stored energy in different types of nuts is converted to heat. The amount of heat generated will allow you to determine how many calories the nuts contain.

# So What Seems to Be the Problem?

All foods contain calories, but, as you know, some foods contain more than others. A 4-ounce candy bar, for instance, contains more calories than four ounces of carrots.

The flip side to the candy bar and carrot comparison, though, is that a candy bar—because of the calories it contains—contains more *energy* than carrots (you can relate that information to your mom the next time she tries to give you carrots as a snack).

You get energy from the foods you eat, because stored energy in those foods is released in a chemical reaction as your body metabolizes it.

**Basic Elements**

Although there are different types of **energy,** the most common form is defined as "the capacity to do work and overcome resistance."

**Scientific Surprise**

Animals that hibernate in the winter—such as a polar bear—have an instinct to store an abundance of calories before they go into hibernation. This gives them the energy they need during the months they go without food.

The energy contained in foods is discussed in terms of heat because that is what is produced when our food reacts with the oxygen (carried by red blood cells) in our bodies. The complete combustion process also liberates water and carbon dioxide. In our bodies, this combustion reaction is known as metabolism.

The problem is, when you consume more calories and their resulting energy than you expend, you're taking in more calories than you need. You need a certain number of calories every day to have enough energy to do what you need to do and to maintain a healthy body.

Many people, however, consume more calories than necessary, and the extra calories—the ones that aren't used as energy—get stored as fat.

In this experiment, you'll attempt to solve the problem of determining the calorie content of different types of nuts. This will tell you how much energy is in the nuts.

If you want to use the title of this chapter, "How Much Energy Do Different Types of Nuts Contain?" for the title of your project, go ahead. Or you might consider using one of these titles:

◆ Releasing Stored Energy to Produce Heat

◆ The Heating Power of Common Types of Nuts

After you've got a title in mind you can start thinking about your hypothesis.

# What's the Point?

Understanding how calories are measured, and how many calories are contained in different types of food will allow you to choose foods that will provide enough calories to keep you energized and moving, but not so many that you'll end up with extra fat. Humans, after all, do not hibernate.

The experiment in this chapter measures the energy within various types of nuts.

Obesity is a huge problem (no pun intended) in America today, drawing increasing alarm from the medical community. Some of this problem is caused by the fact that people aren't aware of the calorie content of the foods they eat.

Making yourself more aware of how food energy works, and knowing how many calories are contained in different types of foods, will help you to have a healthier diet and to feel well and energetic.

Food labels list the amount of calories available per gram of food or per serving. The energy from the foods we eat is used to carry out all of the different chemical reactions that occur in our body.

The experiment you'll be conducting will help you understand how stored energy is converted into heat energy, which is measured in calories.

Most heat energy is measured in units called Joules. We use the unit kilocalorie, or the more abbreviated form, simply *calorie*, to describe the amount of energy we obtain from the foods we eat.

Using a 2,000-calorie daily diet as your control, you'll measure the calories in several different kinds of nuts. The nuts will be the variables, and you'll learn what percentage of daily calories they contain. When you finish, you should have a better appreciation of how stored energy is released, and a better idea of what a calorie is.

# What Do You Think Will Happen?

Using a variety of nuts such as peanuts, almonds, cashews, Brazil nuts, pecans, and hazel nuts, you'll calculate during this experiment how many calories are contained in the different varieties.

To do this, you'll burn the different varieties of nuts in a controlled environment, allowing you to see how much heat energy is released. The released energy will be used to heat a specific amount of water.

> **Basic Elements**
>
> Technically, a **calorie** is defined as the amount of heat needed to raise the temperature of one kilogram of water by 1 Celsius degree.

When you're working to formulate your hypothesis, think about the different sizes of the nuts you'll be using. Are some much larger than others? A Brazil nut, for instance, is about twice as large as a hazelnut. Do you think a larger nut will necessarily release more energy? Are some of the nuts more oily than others? Oil is high in calories and contains a lot of stored energy.

Using knowledge you may already have, and a good dose of common sense, come up with a hypothesis about what you think will occur during this experiment. Which nuts do you think will release the most calories in the form of heat?

# Materials You'll Need for This Project

If you're able to do this experiment in your school lab, you'll have a much easier time. The experiment requires science lab equipment, which most people don't have available at home.

The equipment and materials that you'll need are listed below:

- ◆ Different types of nuts, including Brazil nuts, cashews, pecans, hazelnuts, peanuts, almonds
- ◆ A pan balance or electronic balance
- ◆ Wax paper
- ◆ A sewing needle
- ◆ A cork
- ◆ A Celsius thermometer
- ◆ A ring stand

◆ A ring clamp

◆ An S-hook

◆ A soda can

◆ Matches

◆ An insulated cylinder that will serve as a chimney

Be sure to have everything you'll need for the experiment ready before you start. Whether you're conducting the experiment in your school or at home, be sure to have adult supervision and work very carefully.

**Explosion Ahead**

This experiment is best done in a school science laboratory. You must take the following safety precautions: tie your hair back, wear safety goggles, have an adult supervise you, and keep a fire extinguisher readily available. Don't try to do this on your own, or at home unless you have adult supervision.

# Conducting Your Experiment

Before you begin the actual process of burning the different types of nuts, you need to find out how much each one weighs. To avoid having oils from the nuts build up on the balance scale, put a small piece of waxed paper under each nut before you weigh it.

Once the waxed paper is on the scale, set the scale to zero, if possible. If you can't set it to zero, you'll need to subtract the mass of the waxed paper from the mass of the nut and the waxed paper.

The soda can you use in this experiment must be empty, but still have the tab attached to the top. You'll use the tab to hang the can from the S-hook.

Record all your measurements and calculations on a chart similar to the one shown in the next section, "Keeping Track of Your Experiment."

Follow these steps:

1. Measure and record the mass to the nearest 0.01 grams of the first nut to be tested.

2. Insert a sewing needle into a cork base, and carefully place the nut on the tip of the needle.

3. Weigh an empty soda can and record the mass to the nearest 0.1 grams.

4. Fill the soda can about halfway with water.

5. Measure and record the mass of the can and the water to the nearest 0.1 grams.

6. Calculate the mass of the water by subtracting the mass of the can from the mass of the can plus the water.

7. Once your initial measurements are completed, set up your equipment as shown in the illustration, except do not yet place the soda can into the chimney. Do not attach the can to the ring clamp—attach it to the S-hook.

*This is the setup you'll use to measure how many calories of energy are found in various types of nuts.*

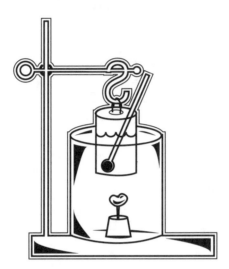

8. Measure and record the temperature of the water in the can to the nearest 0.1 degree Celsius.

9. Light the nut you've put on the needle with a match, and place the chimney over the burning nut.

10. Carefully lower the soda can into the chimney so that it appears as it does in the illustration.

11. Using the thermometer, gently stir the water in the can as the nut burns. This will distribute the heat being transferred from the burning nut to the water more evenly.

12. Allow the nut to burn until the flame goes out. Record the final temperature of the water as soon as the flame has extinguished.

13. Using the equation below, calculate the energy in calories per gram of nut. Knowing both the initial temperature of the water and its highest final temperature allows you to determine how much energy each nut had, and how many calories of energy it contained.

$$\text{Available Energy in calories per gram} = \frac{\text{mass of water} \times \text{increase in temperature}}{\text{mass of nut} \times 1{,}000}$$

14. Repeat the experiment with the same type of nut two more times to give three trials.

15. Repeat steps 1 through 14, using all the other nuts. You will do the entire experiment 18 times if you use all the suggested types of nuts.

# Keeping Track of Your Experiment

Use the following chart, or make your own, similar chart, to keep track of your measurements and observations. When you've got all your data recorded carefully, take a good look at it and compare how well the different kinds of nuts heated the water contained in the can.

| Type of Nut | Mass in Grams | | |
|---|---|---|---|
| | Trial | | |
| | 1 | 2 | 3 |
| Almond | | | |
| Brazil | | | |
| Cashew | | | |
| Hazelnut | | | |
| Peanut | | | |
| Pecan | | | |

*Use this chart to record measurements and observations pertaining to each type of nut used.*

# Putting It All Together

Now that you've conducted this experiment and determined how many calories are contained in each nut, you'll have a better understanding of how calories are measured, and how stored energy becomes heat.

Were you surprised at how well the nuts burned, or how much heat they produced? Did some burn longer than others?

If you compare your body to the equipment you used to conduct this experiment, your body would be the chimney. The oxygen in your body is what allows the food you eat to "burn," or metabolize, just as oxygen in the air fueled the fire as the nuts burned.

# Further Investigation

If you enjoyed this experiment, you could take it a step or two further by using different variables or changing your procedure somewhat. You also could research calorie amounts for nuts and compare your findings with your experiment results. Were you close?

To change your variables, try using different types of food. Some possibilities are avocado, small pieces of meat or fish, small pieces of chicken, both with and without skin, and coconut.

You could even compare the amount of heat given off by a candle to the amount of heat released by the various nuts.

Conduct the same experiment, using a burning candle in place of one of the nuts, and see how effectively the candle heats the water in comparison to the nuts.

Make sure you obtain the mass of the candle before you light it, and, of course, use caution and correct safety procedures when working around an open flame.

## The Least You Need to Know

- Our bodies get energy from foods we eat because stored energy in those foods is released as the food is metabolized.

- Stored energy can be converted into heat energy, which is measured in calories.

- The experiment in this project requires some equipment and necessitates burning, so it's best if you can do it in a science lab in your school.

- Keep accurate and complete information as you conduct this experiment because it involves a lot of measurements that could affect your results.

- Changing the variables or altering the experiment's procedure could be means of exploring this project a little differently.

# Chapter 27

# Other Neat Projects and Activities

## In This Chapter

- Science as a life-changing event
- Fooling your friends with invisible ink
- The incredible bouncing eggs
- Testing your smelly sneakers for bacteria
- Learning about solvents and solutes the easy way

Once somebody gets interested in science—any aspect of science—he or she tends to look at the world a little differently.

Let's say that that someone is you, and you've developed a real interest in biology. You had this great science teacher this year who brought biology alive for you. You're hooked.

Sure, you always liked animals, but now you *really* like animals. And you want to know everything about them. You want to know where they live, how they survive in the winter, what they eat, how they mate, how many offspring they produce, and, well … you want to know it all.

This interest causes you to watch animals closely when you're around them. It prompts you to tune in to animal shows on television, to take out books from the library to learn more about animals, and to search Internet site after Internet site.

Perhaps your interest is contagious, and your family starts to build animal watching into its vacation plans. You begin to see the world in terms of biology and its related disciplines. You are—for your own intents and purposes—a scientist.

It's a proven fact that kids who are exposed to science at an early age in an interesting, innovative manner tend to stay interested as they get older. These generally will be the ones to study science in college and perhaps pursue some aspect of it as a career.

A great way to make science fun and interesting is to experiment and do projects on an informal basis. You don't need to have a science fair looming on the horizon in order to work through a science project.

The next time you and a friend are sitting around trying to think of something to do, suggest that you try one of the projects outlined in this chapter. You'll have a good time, and, who knows, you might even end up looking at the world a little differently.

# Getting the Dirt on the Soil Around You

Dirt is dirt, right? You can plant stuff in it, you can walk over it in your bare feet, or you can play ball on it. If you look carefully, however, you'll not only discover that there are different types of dirt—or soil—but that it contains some very interesting things.

This project doesn't include an actual experiment, although you probably could come up with one if you thought about it hard enough. All you'll do in this project is carefully observe a sample of soil and note your observations. The point of the project is to make you more aware about what soil contains, and how it differs from area to area.

**Standard Procedure**

This project also is fun to do at the beach, using a sand sample instead of soil. You may find many different types of living critters if you take a sand sample that's fairly close to the water.

All you'll need for this project is a sample of soil from the area in which you live, a magnet, a mass balance (a small, electronic scale), and some sieves for sifting the soil. You can make your own sieves from screens of different sizes. Some of the screening should contain fairly large openings, while other openings should be very fine.

Dig up a bucketful or so of soil, and take it someplace where you can spread some out to look at. Take some time and really examine the soil. Do you see anything moving? How does it feel? Does it have a particular odor?

Get your science journal or notebook and write down everything you observe. Think about how the dirt feels in your hands. Is it warm, or cool? Is it a uniform color, or does it contain particles of different colors? Note where you got the soil from, and whether it was easy or difficult to dig up.

Stack your sieves with the largest gauge of screening on the top. If your sieves don't have frames, you can just get someone to hold a piece of screen while you pour dirt onto it. Sift the dirt through the screen into a dishpan or other container below it.

Weigh the particles that remain on each screen, keeping track of which sieve you used and the amount of particles. Use three or four screen sizes, repeating the procedure above for each size.

Save the particles that remain on each screen, and examine them carefully to see how they're different. Do the larger particles have characteristics that the smaller ones do not? Are the particles that collected on each sieve about the same size, or do they vary? Do you recognize any rocks or minerals in the soil?

You can use the magnet to check whether the soil contains any iron. Iron particles will stick to your magnet, and you can weigh them to see if they're heavier than an equal amount of soil without iron.

**Basic Elements**

Soil that is a mix of sand and clay is called **loam**.

Once you've finished examining the soil, add a little bit of water—just enough to moisten it, not turn it into a bucketful of mud.

Then roll some of the moistened dirt between your palms, trying to form it into a ball or disc. If the soil sticks together, that means it contains a lot of clay. If it doesn't stick together at all, it's made up mostly of sand and tiny pieces of rock.

If it sticks together somewhat but is not altogether cooperative, then it's a mixture of sand and clay. Make some notes about how well your soil holds together, and try to figure out what type it is.

If you're interested, you could do some research to learn what type of plants grow best in different types of soil. Or, simply take a walk around your neighborhood, and see what types of plants are prevalent.

Hopefully, once you've done this project, you won't ever again take soil for granted. Think about people who depend on the soil for their livelihoods, such as farmers, tree growers, and landscapers. Try to think of ways in which we can conserve soil, so there will be plenty around for future generations.

# Making Invisible Ink

This simple experiment is great for young kids, but must be conducted with adult supervision because it involves using the kitchen stove.

Invisible ink has fascinated kids for generations. You might think that it would be difficult to make, but it's not.

All you'll need for this experiment is some milk; some white paper; a clean cotton swab or small, clean paintbrush; the kitchen stove; and help from an adult.

All you need to do is, using the brush or cotton swab, write your message with milk on the paper. Don't use a lot of milk and soak the paper. Use just enough to make a thin coating.

Once you've finished writing, ask the adult who's helping you to turn on the kitchen stove to a low heat. Very carefully, using kitchen mitts, hold the paper about a foot above the warm burner. If you don't see results, you may have to make the burner a little hotter.

Do not, however, hold the paper low over the burner. As the paper heats up, your message will become brown and you'll be able to see what was written.

You want to know why? It's because chemical compounds in the milk have a low burning point. When the paper gets warm, those compounds will heat up and turn brown, while the paper will be unaffected.

You also can try this experiment with other liquids, such as orange juice, lemon juice, vinegar, and apple juice. Just be sure to follow all safety rules and have an adult around to help you.

# Making Eggs That Bounce

Did you ever try to bounce an egg? Don't! Not until you read through this experiment, anyway.

This experiment is more fun if you do it with a couple of friends, because you can have a bouncing egg contest when you've finished it. So round up a couple of pals, a dozen eggs or so, and let's get started.

As you know, eggs have shells that you must remove before eating.

If you put eggs in their shells into vinegar, however, the vinegar will do the work of removing the shell for you. That's right. The vinegar will dissolve the eggshell. This happens because vinegar contains an acid called acetic acid, which reacts with the high calcium content of the eggshell.

**Standard Procedure**

Be sure to completely cover the egg with vinegar so that the entire shell dissolves. If there are any pieces of shell left, the eggs won't bounce.

It takes three or four days for the shell to dissolve completely, and you must wait until all the shell is gone before trying to bounce the egg.

All you need to do is put some raw chicken eggs into clear, plastic cups. One egg to a cup, please. This allows you to watch the reaction as it occurs. Completely cover each egg with vinegar, and let them sit in a place where they don't have to be moved.

Check the eggs now and then over the next four days, noting any changes that you see in the shells.

After four days, very carefully, using plastic spoons, remove the eggs from the cups and lay them onto a couple layers of paper towels in order to drain. You'll notice a waxy looking coating on the eggs, but no more shell.

Starting at about 2 inches high, drop an egg onto a tabletop or other surface, and watch it bounce. Working in increments of 1 inch, drop the egg from increasing elevations, having your friends do the same if you conducted this experiment with others.

**Explosion Ahead**

Don't get carried away with your egg drop, because this experiment can get messy. It's a good idea to do the drop outside, or in an area that can be easily cleaned, if necessary. An outdoor picnic table or bench would be perfect.

Whoever has the egg that survives a drop from the highest height, wins.

Don't forget to record your observations as you drop the egg, noting drop heights and how the egg reacts to each drop.

# Do Odor and Bacteria Go Hand-in-Hand?

You know how sneakers get during those hot summer months when you wear them all the time, sometimes without socks? They get kind of, well … stinky.

This is especially true if you get them wet in the creek one day and then forget to put them out in the sun so they dry properly. Damp, dirty sneakers can get really, well … stinky.

Does that smell, though, mean that your sneakers are filled with bacteria and are unsafe to wear?

---

**Scientific Surprise**

Bacteria are everywhere, from mountaintops to the bottom of the oceans. Bacteria, which is the plural for bacterium, have even been found in frozen rocks in Antarctica.

---

In this experiment, you'll test your sneaks against some other, less smelly objects, to see if smell and bacteria might be related.

You'll need to get some petri dishes that contain nutrient agar before you begin your experiment. You can get these at a biological supply company (see Appendix C for the names of some of those companies), or your science teacher might be able to give you a few.

Round up some objects that you want to test, and the most stinky pair of sneakers you can find. Be sure to choose a variety of objects, such as the pencil you use to do your homework, the doorknob of your front door, the inside of your bathroom sink, and so on. Test as many places as you like, as long as you have a petri dish in which to transfer any bacteria you collect.

Be sure to record detailed observations about the objects you're testing, including the following:

- Does the object appear to be clean or dirty?
- Does the object smell bad?
- Where was the object located?

Using a clean cotton swab, wipe the surface of the object you're testing, and then wipe the swab across the surface of the petri dish. Be sure to label each dish, so you know which object the sample was taken from.

Cover the dishes and put them in a warm, dark place where they won't be bothered. Check them twice a day—once in the morning and once at night—for a week, and record your observations about what's happening.

After a week, you'll be able to see which items you tested contained the most bacteria. Were they the dirtiest-looking items? The smelliest ones? Make a chart on which to record your findings, then be sure to dispose of the petri dishes properly. And, just a thought … if your sneakers turn out to be the vacation spot of the world for all kinds of bacteria, you might want to talk to your mom or dad about getting a new pair!

# The Sweetest Experiment

The town of Hershey, Pennsylvania, has a slogan in which it claims to be the sweetest place on earth. That may or may not be true, but one thing is certain. The chocolate that is made in Hershey, Pennsylvania, sure is sweet, and that's part of why you'll love the science experiment outlined in this section.

If you did the experiment in Chapter 9 you already know about solutions and solutes. If you didn't do that experiment or read the chapter, you might want to spend a few minutes to have a look at it.

Briefly, a solute is something that gets dissolved. A solvent, on the other hand, is the substance in which the solute is dissolved.

Do you ever make hot chocolate from the powdered hot chocolate mix? You heat a cup of water or milk in the microwave, and then put the powder into the hot liquid, right? When the powder dissolves, you've got hot chocolate.

**Basic Elements**

The mixture that results when the solvent has completely dissolved in the solute is called a **solution**.

The water or milk in your cup is the solvent, and the powder that you pour into the cup is the solute.

In this, the sweetest of all experiments, you're going to test three methods of dissolving chocolate candy, such as a Hershey's Kiss. All you'll need is three Hershey's Kisses, or a similar, soft candy, your mouth, a clock or watch, and paper and pencil.

The first step of the experiment is to simply place the candy in your mouth and do nothing. Don't move your tongue around in your mouth, and, whatever you do, don't chew. You're letting the saliva in your mouth act as the solvent—the material in which the candy solute will eventually dissolve.

The only real work in this part is that you've got to note the time that the candy goes into your mouth, and the time at which the candy has completely dissolved.

For the second step, put the candy in your mouth, noting the time. This time, you can move your tongue around in your mouth, sort of pushing the candy from side to side. Do not chew. When the candy has completely dissolved, record your time.

With the third piece of candy, record the time that you put it in your mouth, and then go ahead and chew the candy until it's gone. Mark down the time at which the entire piece had dissolved.

It probably will be no surprise to you that the piece of candy you chewed dissolved the fastest, and that the piece that just sat in your mouth took the most time of the three to dissolve.

This is because moving the candy around in your mouth exposed more of the surface area to the solvent (your saliva) than when the candy was just sitting still in your mouth. Just like with the hot chocolate mix, the candy dissolved by spreading out evenly in the solvent.

If you think about making hot chocolate, you'll realize that the process is more successful when you stir the powder in the milk or water instead of just pouring it in and letting it sit there. That's because stirring spreads the solute evenly in the solvent.

So you see, science doesn't have to be at all boring or dry—it can be exciting and lots of fun. Science is a living, always expanding field, and is important to every aspect of our lives.

We all rely on science for our food, our health, and our very lives.

## The Least You Need to Know

- An early interest in science can result in lifetime learning, and perhaps a desire to pursue science as a career.

- Easily scorched chemical compounds in milk make it the perfect material for making your own invisible ink.

- Eggshells dissolve in vinegar and are replaced with a rubbery substance that allows you to be able to bounce the eggs.

- Bacteria is everywhere in the world, but may be found in concentrated numbers in your old, smelly sneakers or other shoes.

- Solutes dissolve best when moved about or stirred within the solvent because more of the surface area is exposed.

# Part 6

# Preparing Your Project for the Science Fair

Once you've finished your experiment, recorded and analyzed your results, and completed any charts and graphs you might be using, there's still a lot to do to get ready for the science fair.

In Chapters 28 through 30, you'll learn how to create a presentation that will showcase your project and impress the judges. The chapters cover some practical matters such as transporting your display and what to do with it when the fair is over.

You've worked hard on your project, so it's important to display it properly and make sure it gets to the fair in tip-top shape. That way, everyone will be able to recognize and appreciate your efforts.

# Chapter 28

# Creating a Great Display

## In This Chapter

- ◆ Getting the right display board
- ◆ Knowing what to include with your display
- ◆ Checking the rules to avoid problems later
- ◆ Making a display that's easy on the eyes
- ◆ Using creativity to make your display stand out
- ◆ Getting your display to the fair safely
- ◆ Knowing what you might need to get your project set up

You can have the most interesting science fair topic in your school, and research your topic until you're blue in the face and the librarian is tired of seeing you show up. You can have the most detailed, well-organized notes of anybody in your class, and use the utmost care when conducting your experiment.

If your display is sloppy and carelessly put together, however, your project won't be recognized as being really great.

Let's face it. Appearance is important in just about every way. We often judge things—and unfortunately, many times people as well—by appearance rather than by content.

Don't think that, just because you've done a good job on other aspects of your science fair project, you can skimp on your display. Even a so-so project can be enhanced by a really great display. In this chapter, we'll cover many aspects of science fair displays, starting with the display board itself. By the time you're finished reading, you should have the knowledge necessary to build a display that will make your project its best.

# Choosing a Display Board

The most basic element of your display is the display board, or backboard. The display board is the background for *almost* everything that will be included in your exhibit.

You can build your own display board, or you can buy one at your local craft or office supply store. Most pre-made display boards are three-sided, cardboard, and designed to stand on a tabletop or counter. It probably will cost you about $5 to buy a display board, and, if you're careful about how you attach materials to it, you might be able to reuse it. Some boards are made from foam, and might cost a little more than their cardboard counterparts.

If you need a sturdier board for some reason, or one that's larger or shaped differently than a pre-made board, you can build your own without too much trouble.

Three sides of a tall cardboard box—cut to the desired size—make a fine display board, as do three pieces of wood, attached with hinges. When deciding what type of board you'll need, keep the following factors in mind:

- How much weight you'll be attaching to the display board.

- How you'll be transporting your display board to the science fair

- Whether or not you'll have time to make one

- How much you have to spend on a display board

- How large you need the board to be

Once you've got your display board, you can begin planning its content and layout.

**Explosion Ahead**

Before you do anything else, read the rules of your science fair and note everything that pertains to the display. Not following the rules can get you disqualified from competition. This should be your very first step in designing and building a display.

**Standard Procedure**

Most science fairs allow exhibits to be a maximum of 48 inches wide, 30 inches deep (front to back) and 108 inches high (including the exhibit table). Generally, these are the maximum sizes allowed, but check to be sure your fair doesn't have different rules.

# What Your Display Board Should Include

Your display board should contain material that sums up your science fair project. A judge or visitor to the fair should be able to look at your display and get a good idea of everything your project entailed.

Your teacher or advisor may have rules concerning what you must include on your board, and even where different components must be placed. If not, however, and you're not sure what your display should include, think about the steps of the scientific method, and the steps you took during the course of your project.

*A neat, attractive display board complements a carefully done science fair project.*

Basically, your display should include the following components:

- ◆ Project title
- ◆ The problem you attempted to solve during the course of the project
- ◆ Hypothesis
- ◆ The experiment (list materials used and describe the procedure)
- ◆ Results
- ◆ Conclusion
- ◆ Future plans

**Standard Procedure** _____

Suggestions are offered in this section for what your display should include, and where different components should be placed within the display. Remember that these are only suggestions. If you have another idea for your display that you think looks terrific, go ahead and use it.

◆ Data

◆ Report, abstract, and journal

These components should be placed on your display board in a logical manner, beginning with the early steps—the problem and the hypothesis. Plan the design so that viewers can follow your signs easily. Let's take a minute to consider each of these components, and where on your display it might belong.

## Project Title

Generally, the title to your project should be prominently displayed, and neatly done. Most fairs do allow signs to be hand-drawn, although you might get a neater product if you do it on a computer.

If you have a three-panel display board, the title of the project probably should be posted at the top of the center panel, unless you have a special reason to put it else-where.

## The Problem

The problem you attempted to solve during the course of your experiment should be placed high on your display board, allowing viewers and judges to see it immediately.

Typically, the top of the left-sided panel is a good place to state your problem.

> **CAUTION**
>
> **Explosion Ahead**
>
> Arranging your display board in a manner that is not easy to follow and understand will result in confusion, and make the viewer have to work too hard to know what you've done. Be sure the display flows in a logical, orderly manner.

## Hypothesis

Depending on how your display is configured, the hypothesis should be posted just below, or across from, the problem. Make sure that it's easy for viewers to follow from one to the other.

Viewers and judges will be unimpressed and quickly lose patience with a display on which they have to search for various components.

## The Experiment

A task that's sometimes difficult is being able to summarize your experiment sufficiently to fit on a single sheet of paper. If you can't do that, it's okay to use a bigger sheet, or to attach two sheets on your display.

Don't be tempted, however, to include every detail of your experiment on your display. Just summarize the procedure—including a list of the materials that you needed for your project.

## Results

As with the experiment, your results should be summarized in a few paragraphs. Remember that your data also will be displayed, so your results need only to be a narrative of what occurred during your experiment.

## Conclusion

Your conclusion needs to be only one or two short paragraphs, summing up what you've learned from your experiment.

You can note in your conclusion whether or not your hypothesis was correct.

## Future Plans

This is an optional component, but judges seem to like it. You'd simply write a brief explanation of how you could expand upon your project, or aspects of the project you would do differently if you were to do it again.

The information you'd present could be something like that included in the "Further Investigation" sections of this book.

## Data

It's very important to include your data with your science fair display. Judges need to be able to examine your data in order to be sure it supports your results and conclusion.

The data would include any charts or graphs you made, and you could include photographs with this part of the display if you've taken some.

**Standard Procedure**

Including photographs gives the judges a much better picture of how your experiment was conducted, possibly resulting in a higher score.

## Report, Abstract, and Journal

If you've done a report or research paper for your project, it should be included, along with your abstract and journal, with your display.

You'll read all about these components in Chapter 30.

Make sure that everything accompanying your display is firmly attached. Anything not posted on the display board—such as your research paper or journal—should be connected to the board with a string or other method.

Now that you know the components of a great display, let's discuss how to make it one that will catch the eyes of the judges.

# Making Your Display Visually Pleasing

While some of you may fall into the category of "artistic type," others probably have a hard time drawing a straight line. Most of us are not blessed with the talents of a great artist. That doesn't mean, however, that your science fair display can't be visually pleasing and impressive looking when the judges come around.

## Using Color

There's nothing that says your display has to be boring. Feel free to use color, either as a background, or in your lettering.

> **Scientific Surprise**
>
> Color is good, but good old black and white, when presented a little differently, can be very dramatic. A display that won high marks in a local fair used a black background with white letters for the title and captions.

Do be careful, however, of using so many different colors—or colors that are so bright—that it's all a judge will see when he or she views the display.

You know that some colors look better together than others, so keep color combinations in mind as you design your display. If you choose a dark color for your display board, consider light colors for the signs you'll post in order to achieve a contrast.

> **Standard Procedure** _____
>
> Check out your local craft or office supply store for pre-cut, adhesive letters. If applied neatly, they make a great presentation on a display board.

## Neatness

You might have a lot of leeway in how your display get sets up, but sloppiness is not an option. Nothing turns off a judge faster than a display that appears to have been thrown together at the last minute with little or no care taken in its appearance.

Be especially careful about neatness if you hand-letter your signs instead of using a computer. Printing by hand is fine, but be sure that your letters are an even size and don't slant upward or downward on the display board.

## Using Your Space Wisely

You'll only have a certain amount of space on your display board, which means it's important that you resist cramming it with every bit of information that you have.

A cluttered display board is confusing and hard to read. Space out the different components of your display neatly, leaving room between each one and placing them evenly on the board.

Make sure that your lettering is easy to read, and that no one part of the display overwhelms another.

# Making Your Display Stand Out

Large science fairs can include hundreds of displays. If yours looks just like all the others, it may not get the attention that it deserves.

There are all sorts of ways to make your display stand out. Color certainly is one way, as is a different sort of lettering style than is normally used. Your computer contains different fonts, or type styles. Experiment with some of them to see what you like and what looks nice on your display board.

If it's permitted, you could consider outlining your display board with a string of small white lights. Or, you could arrange it differently than the "normal" project.

Whatever you do, be sure that it's neat and carefully put together. And remember, it's a fine line sometimes between noticeable and gaudy.

**Explosion Ahead**

Don't work so hard to make your project stand out that it ends up being gaudy and overdone. Too-bright colors or other distasteful attention getters can do more harm to your project than good. You want it to catch the collective eye of the judges, not make them laugh.

# Using Models to Make Your Display Better (Sometimes)

If there is a model that's appropriate to your science fair project, including it can be a great way to enhance your display. That only pertains, however, if the model is applicable and done well.

Including a poorly designed or built model just for the sake of having it will not improve the overall quality of your display.

Be sure that your model isn't so large that it overwhelms the rest of the display. And, try to make it blend in with the rest of the display by matching the colors.

# Safety Rules

Rules vary greatly from fair to fair, depending primarily on the level of competition. Rules that apply to the Intel International Science and Engineering Fair, for instance, are radically different than those for an elementary school–level fair.

Most safety rules, however, are just common sense. All fairs (perhaps with a few exceptions) prohibit students from using metal nails, tacks, or staples to attach live wires to a display board. Any electrical-related part of your display is sure to be scrutinized for safety, so be sure you know what's allowed, and work carefully to make sure it's done properly.

Higher-level projects that involve the use of blood or tissue fall under stringent safety regulations, as do those projects that involve animals. Some rules transcend local fair regulations and are governed by state or even federal rules, such as those concerning hazardous or controlled substances.

## Knowing What the Science Fair Allows

In addition to safety rules, there will be general rules pertaining to your science fair, as well. As mentioned, the size of your project will be a consideration, as will rules pertaining to drop-off and pick-up times, viewing times, and so forth.

**Standard Procedure**

It seems that science fairs are constantly adding rules, and the higher-level the fair, the more rules there are likely to be. Breaking even one rule can disqualify your project.

It's very important to get a copy of the rules that apply to your particular science fair, and to become familiar with them, before you even start working on your project.

Claiming that you didn't know about or understand a rule won't help you if you show up at the fair with a project that doesn't meet all regulations.

## Transporting Your Project to the Fair

You've probably worked for weeks on your science fair project, and possibly longer. When the time comes to transport it to the fair, you definitely want to make sure that it gets there safely.

Your best bet is to ask a parent or other adult to help you by driving it to the school or other location at which the fair is held. If you live in a city and use public transportation, it might be easier to transport the pieces of your project and assemble it at the fair.

## Size Counts

If your project is too large or unwieldy, you'll have a difficult time transporting it. And, the more unwieldy it is, the greater the chance is that the project could be dropped or otherwise damaged.

While you want your display to be eye-catching and attractive, there's no gain in making it bigger than it should be or needs to be.

## Getting It There in One Piece

If your project contains a lot of small pieces, you may want to consider transporting them separately and attaching them to the display once it's set up at the fair.

Even if your display is fairly simple, take the utmost care when moving it from one location to another. Anything that could spill or break should be protected and very carefully transported.

> **Standard Procedure**
>
> If you're transferring parts separately and assembling your project at the fair, make a list of everything you might possibly need and check it off before leaving the house.

# Getting Your Project Set Up and Ready

Most projects require some setting up, and you've got to be sure that you're prepared to do that once you reach the fair.

If your project requires assembly at the fair—that is, you can't just set up your display board and place your report and journal on the table in front of it—you'll need to make sure you have all the materials you need.

Some things to consider when deciding whether you'll need to get your project up and ready are listed below.

◆ Tools or equipment to assemble the display

◆ Electrical extension cord

- Three-pronged plug

- Extra letters, glue and tape

- String to attach journal and report to display

- Stapler, scissors, tacks

- Extra markers and paint

Also, be sure that you understand when displays must be set up and taken down, and how long you'll have to do so.

There's a lot to consider when planning, designing, and building a science fair display, but with some careful thought and some hard work, you should be able to create a great display.

Don't be afraid to ask a parent, teacher, or older student for their opinions and advice, and leave yourself enough time to finish everything you need to do.

Above all, have fun with your display, and let it reflect you and the project you've chosen. Science isn't all serious business, you know. It's a living, fascinating field, and your display can reflect those same qualities.

## The Least You Need to Know

- You can buy a standard-made display board, or make one that's customized to your needs.

- Your science fair display should include a title; summations of your problem, hypothesis, experiment, results, conclusion, future plans; data; and your report, abstract, and journal.

- Be sure to get a full set of science fair rules before starting your display so you know if components must be positioned on the board in a certain manner, and other rules that relate to your project.

- Use creativity, color, and neatness to make your display visually appealing.

- A model can enhance your project, but only if it's applicable to the topic and has been neatly and properly constructed.

- Take care when transporting your project to the fair, and be sure you have everything necessary to get it set up and ready for the judges.

# Chapter 29

# Knowing What the Judges Are Looking For

## In This Chapter

- ◆ An overview of judging standards
- ◆ The scientific method strikes again
- ◆ Paying attention to neatness and clarity
- ◆ Comparing your display to others
- ◆ Assuring that your work is presented clearly
- ◆ Preparing to meet the judges

In Chapter 1, you read a little about how fairs are judged. You learned that judges look for certain qualities in a project, and evaluate a project based on several standards.

In this chapter, we'll take a closer look at the qualities judges are likely to be seeking when they evaluate your project. We'll discuss what's most important about your project and display, and take a critical look at how well you've used the scientific method.

Remember that, generally, judges start out by assuming each project they look at is average. Then they either add or deduct points, depending on how well your project measures up against their standards.

It's important to remember, also, that judges aren't nasty people who are out to get you, or hand you a lousy score, or embarrass you in front of your entire class and family. Generally, judges are selected from your community, where they work as engineers, chemists, doctors, psychologists, and in other scientific areas.

They're busy people, but have taken the time to judge your science fair because they're interested in young people, and want to encourage young people to get interested in science, or to pursue their interests. Many of the same folks who agree to judge a science fair would probably be willing to serve as a mentor to a student who is seriously interested in a career in science.

So you see, you really should be grateful to the judges, even if they don't tell you that your project is the best one seen since Albert Einstein was in school. If you've worked carefully, adhered to the scientific method, and done your best, you have nothing to fear from the judges.

Now let's see what judges look for in a science fair project, paying special attention to how well you've employed the scientific method.

# Basic Standards of Judging

If you'll think back to Chapter 1, you may recall five general areas to which judges look when evaluating a science fair project. Just for a review, those categories are listed below, with a brief explanation of each.

♦ **Creativity.** Have you exhibited a sense of curiosity when choosing your topic? Is your display creatively designed and exhibited? Were you creative in how you went about performing your experiment and collecting results? Is there evidence that you thought "outside the box" in a creative fashion?

**Standard Procedure**

These are basic categories that most judges use as their main criteria for evaluating projects. You should, however, receive a paper from your teacher or advisor, telling you what the judges in your particular fair will be looking for.

♦ **Scientific thought.** Did you use the scientific method, including stating the problem, using appropriate materials, proper procedure, observation, and conclusion? Are your ideas original and do they contain scientific value? Did you adequately investigate your topic? Did you gather data in an orderly, scientific manner? Did you use proper experimental procedures?

*A checklist of the five general areas judges consider when evaluating a science fair project.*

♦ **Thoroughness.** Is all information presented in an accurate manner? Did you collect sufficient data? Does your project indicate that you have a full understanding of your chosen topic? Have you included adequate documentation of your work?

♦ **Skill.** Is your display neat and durable, and does it reflect your own work and not your dad's? Does your project meet all safety standards and other fair guidelines?

♦ **Communication.** Is your project self-explanatory and easily understandable to the average person? Have you used correct spelling and grammar? Do your signs, lettering, and diagrams enhance the display or result in it being cluttered and confusing? Is your project logical, clear, and complete?

Once you've reviewed these basic categories, we'll take a closer look at the most important qualities the judges will be looking for when they evaluate your project.

# What's Your Creativity Level?

Some people are just more creative than others, there's no question about it. You know who they are. The kids who not only star in the class plays, but write and direct them, as well. Those who turn a simple art project into something worthy of display in New York City's Metropolitan Museum of Art, or come up with previously unheard of theories to explain the disappearance of the dinosaurs.

If you're lucky enough to be one of those kids, go for it. Be sure to incorporate your natural creativity into every aspect of your science fair project, from choosing your topic to building your display.

CAUTION

**Explosion Ahead**

Creativity is a good thing, but be sure that you stay within the guidelines of your fair. If not, your creative streak could end up costing you points, or even getting your project disqualified.

If you're like most folks, though, and only of average creativity, there are some standards you'll need to keep in mind when working through your project.

While judges will evaluate the creativity of your display, they most likely will be even more interested in how creatively you designed and worked through the steps of your project. If you followed the steps of one of the experiments spelled out in this book, you didn't really need to design an experiment, as you would if you were starting from scratch.

If you exhibited creativity in your work, however, you might have thought of a way to vary the experiment that was explained, or incorporated some of the hints given in each chapter on how to take your experiment a step or two further.

Maybe you came up with a particularly creative idea for a project—one that has never before been done in your school. Guaranteed that judges will be more impressed by a well-done project on an uncommon topic than by a well-done project for which they've seen 50 others just like it.

# How Well Did You Use the Scientific Method?

Just as the scientific method is the backbone to your experiment—to your entire science project, really—it also is the center of what judges will be looking for. There's just no escaping the scientific method when it comes to science fairs.

## Step-by-Step

Take a few minutes to recall the five steps of the scientific method. They are:

- Identify a problem
- Research the problem
- Formulate a hypothesis
- Conduct an experiment
- Reach a conclusion

Now think carefully about your science fair project—one step at a time.

After you selected your topic, did you identify a problem that you would attempt to solve? And did you record that problem in your journal? Perhaps you were attempting to find out the best liquid in which to dissolve salt and sugar, or if the pH factor of various liquids affects the growth rate of plants.

If you didn't specifically identify a problem that you planned to work to solve, you did not strictly adhere to the scientific method.

Now think about the research you did in order to help formulate your hypothesis. This research should have been specific to the problem you stated, and conducted with the purpose of providing adequate knowledge and information on which to base your hypothesis.

**Standard Procedure**

Remember that any previous knowledge you have also is helpful when formulating a hypothesis. That knowledge, combined with some research, can give you a good foundation on which to base your guess.

If you failed to conduct this type of research, or to properly document it, you could end up losing points when judges look to see how well you adhered to the scientific method.

You might have a good idea as to the results of your experiment, but if you don't state a hypothesis, you've ignored an important part of the scientific method. Your hypothesis should be clearly stated—in writing—as part of your science fair project.

As you know, conducting your experiment is the meat and potatoes of the scientific method, and of your entire science fair project, as well. Check carefully to be sure that you worked through your experiment in a logical, methodical manner, that you repeated the experiment if necessary, and that you documented all your observations in a journal.

Once you completed your experiment, recorded your results, and analyzed your data, did you follow through to reach a conclusion that either supported or proved wrong your hypothesis? This last step of the scientific method is very important, and your project isn't really complete until it's been done.

## Is Your Whole Project Covered?

Take a little time to closely examine your science fair project, and think about whether you've worked logically and scientifically throughout.

Even parts of your project that don't apply strictly to the five steps of the scientific method should be orderly, neat, documented, and logical.

Judges will look to see whether your project was carefully planned, if it demonstrates that you fully understand the problem you attempted to solve, and if your experiment was a logical means of solving the problem.

They also will want to assure that you've recorded all your observations, that all information presented is accurate, and that you have sufficient data.

Going through your project step-by-step will help to assure that your work is sound, and go a long way toward impressing the judges.

# How Does Your Display Measure Up?

While your use of the scientific method and the overall scientific value of your project may be the primary standards of judging, they are not the only factors judges will be looking at.

### Standard Procedure

It's always a good idea to get a couple of different people to look at your display and offer advice or constructive criticism. If your sister is a great art student or your brother's friend a straight-A science student, for instance, they're likely to have an idea or two that could improve the quality of your display.

### Standard Procedure

Remember that a display that's a little out of the ordinary will get some attention that others may not. If your display looks just like all the others, it may not attract the attention of the judges, even if it's a really great project.

Chapter 28 explained in detail how to make your display visually pleasing, noticeable, and how to keep it safe during transport.

Suffice it to say here that it's worth a final look at your display to make sure it's neat, self-explanatory, and attractive. Ask yourself the following questions as you view the display:

- ◆ Are any signs you've attached displayed neatly?

- ◆ Are they straight?

- ◆ Is the lettering on them neat and attractive?

- ◆ Are the signs arranged in a logical order, working to explain the step-by-step process you used in your experiment?

- ◆ Is your display pleasing to look at, or cluttered and confusing?

- ◆ Have you included some color?

- ◆ Have you included photographs to help you illustrate your experiment or other parts of your project?

◆ Can the average person look at your display and understand the purpose of your project and how it was carried out?

◆ Does your display adhere to all of the guidelines of your science fair?

If you're satisfied with your answers to the questions above, you should be ready to head out to your science fair, armed with a great display.

# How You Say It

A science fair project is only as good as the level at which it's communicated. A scientist may conduct a groundbreaking experiment and achieve astonishing results, but if he or she can't effectively communicate the information pertaining to it, the work may pass by largely unnoticed.

Judges are likely to check out your notebook or journal to see how your information is presented. Keep it concise and factual. If you speculate or offer your own opinion, make sure it's clearly written that the information presented is so.

You will not win points from judges with a notebook that is cluttered, sloppy, or unreadable.

# Don't Forget to Document

Documentation can be tedious, as you read in Chapter 3. Regardless, it's a vital part of a science fair project, and an area at which judges will look closely.

If you've obtained information from any source, you must document where it came from.

Remember that credit must be given each and every time you:

◆ Use another person's idea, opinion, or theory

◆ Use any facts, statistics, graphs, drawings, or other pieces of information in any form that are not common knowledge

◆ Quote another person's spoken or written words

◆ Paraphrase (put into your own words) another person's spoken or written words

You also must document your own observations, recording when and under what circumstances they were made and noted.

# Preparing to Meet the Judges

At some science fairs, but not all, judges walk through the displays while students are in attendance and ask questions of participants.

Not all science fairs include this feature, but you should be prepared in the event that you'll need to discuss your project with the judges.

**Standard Procedure**

Anticipating questions from judges at a science fair is similar to anticipating questions you might face when interviewing with a college admissions counselor. It prepares you, although not perfectly, for what may occur.

These question-and-answer sessions normally are rather casual and informal, unless a formal interview procedure is in place. Generally, the judges will just ask a few questions about your project, and then move on to the next table.

It's a good idea to anticipate some questions that judges would be likely to ask, just in case you find yourself with an interested judge at your display, wanting to know where you purchased your supplies and how you ever came up with such a clever idea for a topic.

*Be prepared to talk about your project and answer any questions a judge may have during the science fair.*

Let's consider some of the types of questions judges would be likely to ask:

◆ "How did you choose your topic?" This may include questions about research you conducted when thinking about different topics, or whether your project is one that you continued from another year.

◆ "How does your project adhere to the scientific method?" If you remember the method's five steps, you should be able to walk the judge through exactly how you applied them to your project.

◆ "What were your controls and variables?" If you've taken time to assess your project as suggested earlier in this chapter, you should have no trouble answering this type of question if asked.

◆ "How was your experiment conducted?" If asked this question, you may want to refer to your notebook or journal in order to give the judge a complete explanation. Some experiments have a great many steps, and it would be easy to leave one or two out.

**CAUTION**

**Explosion Ahead**

If you don't understand a question a judge asks you, ask him or her to repeat it. If you still don't understand it, politely tell the judge you're unsure of what he or she means. Trying to answer a question that you don't understand will make you sound like you don't know what you're talking about.

◆ "Where did you get your materials and supplies?" This type of question should be fairly easy to answer, although you could consult with your notebook, if necessary.

◆ "Did you conduct your experiment more than once?" If you did, you'll want to explain any concerns that may have prompted you to repeat the experiment.

◆ "How do you interpret your results?" Your answer to this question would be based on your analysis of your data. Be sure you'd be able to explain how you drew your conclusions, and be ready to produce your data, if necessary.

◆ "What were your primary sources of information during your research?" Your sources will be noted and documented in your notebook or journal, but it's a good idea to be able to name two or three of the best ones if you're asked.

◆ "Do you plan to expand your project or use it as a basis for any future research or project?" How you answer that one is up to you. If you're not sure, simply say that you may consider doing so in the future.

Most judges are truly interested in talking with you about your project. Just remember to speak clearly and slowly, look the judge in the eye, and offer your hand if he or she makes a move to shake hands.

Try not to appear overly nervous, and remember that there are a lot of projects for each judge to consider. He or she won't be with you for very long, so say what you want to say while you have the opportunity to do so.

## The Least You Need to Know

◆ Basically, judges consider creativity, scientific thought, thoroughness, skill, and communication when evaluating a science fair project.

◆ Judges will look closely to see if your project adheres to all the rules of the scientific method.

◆ Your display should be neat, arranged in a logical way, and clear enough for the average person to understand easily.

◆ A display that looks different and stands out will get the attention of the judges, as long as it adheres to all rules and restrictions.

◆ Judges sometimes ask questions of science fair participants, making it important for you to be prepared and composed.

# 30

# Science Fair Project Papers and Reports

## In This Chapter

- ◆ Writing an effective abstract
- ◆ The basics of writing a research paper
- ◆ Using a journal to document information and stay organized
- ◆ Understanding the purpose of abstracts, research papers, and journals
- ◆ Using papers and reports to impress the judges

As with many endeavors, your science fair project won't be complete until the paperwork is done.

There has been a lot of discussion throughout this book about researching topics, taking notes, keeping journals, documenting your research, and presenting information in other ways. All of these are very important tasks that will significantly impact your project.

Keeping organized, neat notes throughout your project will help you to keep track of what you've done, what needs to be done, and what you've observed. Even if you've got a super memory, it's nearly impossible to remember every detail of an experiment or exact observation.

In this chapter, we'll recap some of the topics already mentioned earlier in the book. We'll go over some of the basics of researching, and give you additional pointers on writing research papers, abstracts, and journals. You'll learn how to use these works effectively as part of your display.

# The Abstract

An *abstract* is a summary of a piece of work. It boils down a lot of information into just a few sentences, pulling out the essence of all the work presented.

The abstract you'll write to accompany your science fair project must tell viewers—including the judges—exactly what your project is about. For that reason, it's a very important piece of writing, and not always an easy one.

Many people find it easier to sit down and write three or four pages of copy than to capture the essence of a work in a few sentences.

---

**Basic Elements**

An **abstract** is a written summary of a piece of work that in a few sentences conveys the essence of the project.

---

**Standard Procedure**

Writing an abstract for your science fair project is good practice for later in your education. High school teachers are likely to require one to accompany a research paper, and an abstract is an extremely important component of a master's or doctoral thesis.

---

In order to do that, you'll need to identify the most important aspects of your project. This may require some careful thought and analysis. It is essential, however, that your abstract contains only the most significant information pertaining to your science fair project.

Your abstract should include the following:

- The purpose of your science fair project
- The method that you used (a very brief description of your experiment)
- Your results
- Your conclusions
- A recommendation (if you have one)

Chapter 10 described a project that compares the rate at which paper and various types of plastic trash bags degrade when they're buried in the ground.

An abstract that could be used to summarize that project is written here. Take a quick look at the chapter, and then look at how the abstract includes all the points previously listed.

---

**Sample Abstract**

The purpose of this project is to compare the rate at which paper and various types of plastic trash bags degrade when buried in dirt. To achieve this, I completely buried a paper bag and three different types of plastic trash bags for three months. All of the bags had the same contents. During the time they were buried, the paper bag decomposed almost completely. All of the plastic bags, however, remained intact. I recommend that we consider switching from plastic to paper trash bags in order to reduce the amount of plastic that is filling up our landfills.

---

Notice that this sample abstract tells the purpose of the project, very briefly describes the experiment and results, and also contains a recommendation. The entire abstract, however, is less than 100 words.

Some tips to keep in mind as you begin writing your abstract are as follows:

◆ An abstract should not contain any new information, only a condensed version of existing information.

◆ An abstract must be easily readable and understandable.

◆ Information contained within an abstract should be presented in chronological order. For instance, you should state the purpose of your experiment (the problem you hoped to solve) before divulging the results.

◆ An abstract must not ramble or include extraneous information. It should be short and to the point.

◆ Information presented in an abstract should flow logically, with good transitions used to move from one topic to another.

◆ As with any written work, an abstract should be neatly done, and contain no spelling or grammatical errors.

# The Research Paper

You may or may not be required to complete a research report or research paper as part of your project. Some fairs require this, while others do not.

If you do, be sure that you've carefully read Chapter 3, which provides lots of information about researching methods, documenting your work, and taking good notes.

There are several types of research papers.

♦ Some research papers are simply reports on an assigned topic. This type of paper is valuable, because it requires students to conduct and document research, sort through material to determine relevance and importance, and to organize material.

**Standard Procedure**

Your research paper, if required, will report on the initial research you did in order to formulate a hypothesis. The research will provide background information about your topic, and a basis on which to base your hypothesis. This research also will help you better understand the results of your

**Explosion Ahead**

Beware of over-researching a topic. At this level of study, too much information can become cumbersome and make it difficult for you to sort out what is relevant and what isn't. Adequate research is necessary, but don't feel that you need to learn every single fact relating to the problem you're trying to solve.

♦ Another type of research paper is argumentative. With this type of paper, students take a position on an issue and research to find information that supports that position. A major goal of this type of research paper is to teach students to anticipate and refute arguments that others will make concerning their topic.

♦ Another basic type of research paper is one that deals with a particular problem found in an area of study. In this type of research paper, students state the problem and supply information that may be helpful in solving the problem. They then attempt to find an answer to the problem.

As you've probably guessed, if a research paper is required as part of your science fair project, you'll follow the third model, which includes information geared toward solving a problem. The problem, of course, is the question you're attempting to answer in your project.

If your project is modeled after Chapter 17, for instance, your research will lead you to information that explains exactly what DNA is, what it does, and how it varies from species to species. With some diligent research, you'll learn all kinds of information about DNA.

Before you begin researching your topic, take a few minutes to think about the main parts of a research paper. It should include:

♦ A statement of the problem you'll be attempting to solve. Example: *Is the DNA of a Cow Different from That of a Chicken?*

◆ A thesis sentence. A thesis sentence sums up the contents of your paper. Example: *The DNA of a cow is significantly different from that of a chicken.* You'll need to complete your research first in order to be able to write a thesis sentence.

◆ The information you've gathered during your research that relates to your problem and thesis sentence. Remember to include your sources.

◆ A conclusion, based on your research. Your conclusion will be a summary of what you learned, not just a sentence stating your hypothesis.

Research papers at the college level and above can be extremely complicated and involved. At this level, however, a good research paper does not need to be overly complex, as long as it's carefully done, well planned, and contains relevant information.

# The Journal

The journal you keep during the course of your science fair project is extremely important. Your journal is a diary of what your project entails, the procedures you undertake, and your observations. Your journal is primarily for your use, although a judge may want to have a look at it, as well.

You've read in previous chapters about the type of information that should, or could, be included in your journal. Let's have another look, however, just for review.

◆ Research notes. Even if you don't have to do a research paper, you'll still need to conduct research. Using your journal to keep track of your sources can help you to stay organized.

◆ A clear statement of the problem you'll attempt to solve during the course of your project.

◆ A clear statement of your hypothesis.

◆ If you're designing your own experiment, your plans should be included in your journal. Even if you're not designing your own experiment, you should note your control and variables, and how you plan to conduct the experiment.

**Standard Procedure**

Your journal can be whatever form you choose. Some common types include three-ring binders, spiral-bound notebooks, and special books designed for use as journals. Just make sure that your journal is self-contained and in a format where papers won't be lost.

♦ Materials and supplies that you need for the experiment. You also may want to include where you purchased supplies, how much they cost, and so forth. This information may be useful if you do another science fair project in the future.

♦ The steps of your experiment. Whether or not you design your own experiment, you should record in your journal what you did.

**Standard Procedure**

If you're unsure of whether or not to include a particular piece of information in your journal, go ahead. You should be able to look back on your journal long after your science fair project has been completed, and read everything you did, the materials you used, what you observed, and so forth.

♦ Your observations. This is perhaps the most important part of your journal. You should never assume that you'll recall your observations. Always write them down. Use as much detail as possible when noting what you see, including measurements, information about color, and so forth. The more you write down about what you see, the more information you'll have from which to draw your conclusions.

♦ Rough drafts of tables, graphs and charts. You may need to design some of your own charts and graphs on which to record information. If so, you can plan out designs for them in your journal.

♦ Your analysis. Jot down the steps you used while analyzing your data.

♦ Your conclusion. Even though your conclusion will be posted on your display, be sure to include it in your journal. Your journal should be a record of everything you've done during the course of your project.

As with keeping any type of journal, it's important that you be faithful to it. Don't assume that you'll remember to record an observation you've just made in a day or two because you don't feel like taking the time or you're busy doing something else.

While your journal is a tool for you to use, it won't be beneficial if you can't read it or understand what you've written. Take your time when recording information so that it will be useful to you later on. It's a really good idea to date all your journal entries. And it's absolutely necessary to date—and even include the time—of the observations you make during the course of your experiment.

# Why They're Included in Your Display

Your display, as you read in Chapter 28, needs to clearly show viewers what your science fair project is all about. Judges and onlookers at a science fair view a lot of projects in a relatively short time, and they don't want to have to take a lot of time trying to figure out the contents of any one display.

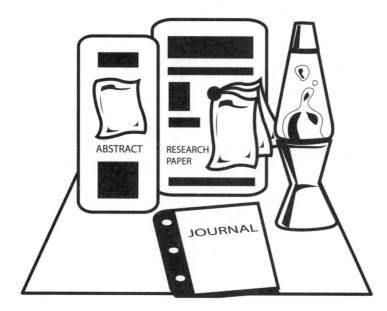

*Your display should include an abstract, your research paper, and your journal.*

Including your abstract, research paper, and journal with your display helps viewers and judges to get a better, more complete idea of what your project is about. The abstract states the essence of your project and tells viewers at a glance what occurred during your experiment.

This can be very helpful in allowing a viewer who just wants to take a quick look at what you've done to understand the purpose and outcome of your project quickly.

**Standard Procedure**

If you leave your journal and research paper sitting out with your display, consider attaching them somehow to the display board so you don't lose them.

The research paper provides background information that relates to the problem you're attempting to solve. It can help judges to understand how you reached your hypothesis, and show them the sources from which you obtained information.

The journal, which provides a place for you to record your thoughts, observations, and other pertinent information about your project, can also be an important source of information for judges.

Let's say that a judge is wondering why you conducted your experiment five times instead of just two or three. Being able to check your journal and read that you had a problem regulating the heat in the room in which you were attempting to germinate bean seeds will let the judge know exactly why you repeated the experiment so many times.

# How They Can Make Your Project a Winner

Your abstract, research paper, and journal, if well done and neatly presented, can be very impressive and useful to the judges at your science fair.

While the most attractive projects tend to get a lot of attention from viewers during a fair, those that have information and research to back them up will get the attention of the judges.

These papers—the abstract, research paper, and journal—help to assure judges that your work is your own, that you fully understand what you've done, and that you could repeat the experiment, if necessary. Those qualities, to a judge, are more important than a nice background color on your display.

## The Least You Need to Know

- An abstract should summarize your project, highlighting the most important points.

- A research paper may, or may not, be required as part of your science fair project.

- Your research paper would include an explanation of a specific problem and information that can support a solution to the problem.

- Your journal is an important tool for keeping yourself organized, as well as for documenting information.

- Abstracts, research papers, and journals are important pieces of a science fair project because they inform viewers and judges about what you've done, and provide you with a lasting record of how your project was conducted.

- Papers and reports are more substantive than a flashy display, and can go a long way in impressing science fair judges.

# Appendix A

# Glossary

**abstract**  A written summary of a piece of work that in a few sentences conveys the essence of the project.

**air pressure**  The push that air has against all surfaces it touches.

**anion**  A negative ion, formed when an atom gains electrons.

**atom**  The smallest particle of an element.

**bar graph**  A type of graph used to show relationships between groups, which is beneficial for showing large differences.

**bibliography**  A list of all the sources from which you obtained information for your research paper, including books, online sources, magazines, government publications, newspapers, journals, and reference materials.

**biodegradable**  A substance that is capable of decomposing under natural conditions.

**biology**  The area of science that deals with various aspects of plant life, animal life, and human life; also deals with the origin of life, its distribution, physiology, development, and habits.

**botanist**  A scientist who studies plants and related subjects.

**botany**  The area of science that deals with the structure of plants, the functions of their parts, the conditions in which they grow, and the language used to describe and classify them.

**caffeine**   A substance found naturally in the leaves, fruits, and seeds of more than 60 kinds of plants, including tea leaves, coffee, kola nuts, and cocoa beans.

**calorie**   The amount of heat needed to raise the temperature of one kilogram of water by one Celsius degree.

**cation**   A positive ion, formed when an atom loses electrons.

**chemical composition**   The materials an object is made of.

**chemical reaction**   A chemical change that produces new and different substances.

**chemistry**   The area of science dealing with the composition and properties of substances, and with the reactions by which substances are produced from or converted into other substances.

**chromatography**   The process of separating mixed color compounds.

**common knowledge**   Facts that can be obtained from a wide variety of sources and are widely recognized as true.

**concentrated**   The state of a solution when a lot of solute has been dissolved in the solvent, but there is still some space between the molecules.

**control**   A factor that remains the same throughout an experiment.

**convection**   The transfer of heat by the movement of currents within the heated material.

**cross-pollination**   The process in which pollen is exchanged from one flower to another, either by natural methods such as butterflies, bees, and wind, or by human means of altering plants.

**density**   The ratio of the mass of an object to its volume.

**dew point**   The temperature at which air becomes saturated with water vapor.

**deoxyribonucleic acid (DNA)**   A substance found in every cell of most living organisms that determines inheritance of eye and hair color, height, stature, and so forth.

**electromagnetic radiation**   Energy that travels in transverse waves and is measured in wavelengths, including forms of energy ranging from visible light to radio waves, microwaves, and gamma rays.

**electron**   A tiny, negatively charged particle that orbits around the nucleus of an atom.

**electroscope**   A device that detects electrical charge.

**energy**   The capacity to do work and overcome resistance.

**environment**    All the conditions, circumstances, and influences that surround and affect the development of an organism.

**evaporation**    The process that occurs when the molecules of a substance in a liquid state absorb enough heat energy to cause them to vaporize into a gas.

**Fleming, Alexander**    The scientist who in 1929 discovered that the mold, penicillin, had the ability to stop the growth of a colony of germs placed into the same petri dish.

**high-density polyethylene (HDPE)**    A type of opaque plastic which is hard to semi-flexible, with a waxy surface.

**horticultural cloning**    The process of propagating plants by vegetative methods.

**hybrid plant**    The offspring of two varieties or species of plants.

**hydroponics**    The practice of growing plants in a water solution to which nutrients have been added.

**hypothesis**    An educated guess predicting the outcome of a science experiment, based on knowledge and research.

**immiscible**    Incapable of mixing or blending, such as in the case of oil and water.

**ion**    An atom or group of atoms containing either a positive or negative electric charge as the result of the loss or gain of electrons.

**Joule**    The most common unit used to measure energy.

**Kids F.A.C.E. (For A Clean Environment)**    A kids' environmental group started with six members in 1989 in Nashville, Tennessee, which has grown to 300,000 members in 15 different countries.

**kinetic energy**    Heat energy which causes molecules within a substance to move.

**loam**    Soil that is a mix of sand and clay.

**low-density polyethylene (LDPE)**    A type of plastic that is soft and flexible, with a waxy surface and low melting point.

**line graph**    A graph containing a "y" axis and an "x" axis that is used to plot ongoing data.

**lysing**    The process of breaking the plasma membrane of a cell.

**mass**    The weight of an object.

**Mendel, Gregor**    A scientist who lived between 1822 and 1884, known as the father of genetics for his work in the areas of genetics and heredity.

**molds**   Microscopic fungi that live on plant or animal matter.

**molecule**   Two or more elements that are chemically combined.

**meniscus**   The curved edge of a liquid in a container such as a graduated cylinder.

**metal alloy**   A mixture of two or more metals, an example of which is steel.

**metallurgist**   An expert in the area of metals who researches, controls, and develops processes used in extracting metals from their ores in order to refine them; also known as a metallurgical engineer.

**metric system**   A system of measurement developed by French scientists in the late 1700s, and used today by the general populations of 99 percent of all the countries in the world.

**Millikan, Robert**   A renowned American physicist and Nobel prize winner, widely known for his oil drop experiment, conducted in the early part of the twentieth century. The experiment is hailed as being simple, yet brilliant.

**momentum**   The mass of a moving body times its velocity. Momentum can be transferred from one object to another.

**monomers**   A repeating unit made up of two or more carbon atoms that are bonded to one another with hydrogen, which is bonded to the carbons.

**natural science**   The areas of science, such as botany, biochemistry, and behavioral and social sciences, that deal with the study of all living things.

**neutron**   A tiny particle, having neither a positive nor negative charge, contained within the nucleus of an atom.

**nucleus**   The center of an atom, containing positively charged protons and negatively charged neutrons.

**oxidation**   The reaction of an object exposed to oxygen.

**pesticide**   Anything that destroys pests or suppresses or alters their life cycles.

**phospholipids**   The two layers of compounds that form the plasma membranes surrounding cells.

**physical science**   A huge area of science encompassing chemistry-related topics such as compounds, molecules, and the chemical elements; physics; electronics; and electricity.

**pie graph**   Also called a circle graph, a pie graph shows the relationship of a part to the whole, and is useful in explaining percentages.

**plagiarism**   The act of taking ideas, facts, quotations, or other information from any source and claiming it as your own by not giving credit to the source

**physical science**   The areas of science that deal with matter, force, and energy, such as chemistry, physics, computer science, and engineering.

**polyethylene terephthalate (PET)**   A clear, tough plastic that's often used as a fiber in fillings for pillows and other items.

**polymer**   A compound made up of repeating units called monomers.

**polypropylene (PP)**   A hard but flexible plastic with a high melting point and ability to withstand solvents.

**polystyrene**   A clear, glossy type of plastic that is rigid and brittle; a variation of polystyrene is foamed plastic, commonly known as Styrofoam.

**polyvinyl chloride (PVC)**   A type of plastic that can be hard and rigid, or flexible and elastic, depending on whether or not it's plasticized.

**protease**   A specific type of enzyme that targets proteins, breaking them down so that they're small enough for the body to use.

**proton**   A tiny particle, having a positive charge, that is contained within the nucleus of an atom.

**pulse point**   The spot on a person's wrist at which you can feel the regular beating in the artery, caused by the contractions of the heart.

**qualitative results**   Results that can only be observed, not accurately measured.

**quantitative results**   Results that can be accurately measured.

**reaction time**   A measure of how fast a person can respond to a situation or stimuli.

**reflexes**   Actions performed by your body independently of the thought process that work to protect you from harmful situations.

**Redi, Francesco**   An Italian physician (1626–1697) credited with being the first scientist ever to use the scientific method.

**rocket engines**   Reaction engines that propel using the forces of action and reaction.

**rust**   The result of the reaction that occurs when metals containing iron react with the oxygen in the air or in water and form a hydrated compound called iron(III) oxide (ferric oxide).

**saturated**   The state of a solution when absolutely no more solute can be dissolved within the solvent.

**scientific method**   A series of steps that serves as a tool used to help solve problems and answer questions in an objective manner.

**search engine**   An electronic procedure that locates web pages that have been identified in advance by a system using key words.

**solute**   The substance—either a solid, liquid, or gas—that gets dissolved in a solvent and forms a solution.

**solution**   A uniform mixture of a solute (usually a solid) dissolved in a solvent (usually a liquid).

**solvent**   A solid, liquid, or gas in which another substance is dissolved in order to form a solution.

**specific gravity**   The ratio of the mass of a solid or liquid to the mass of an equal volume of distilled water at 4 degrees Celsius.

**spontaneous generation**   An ancient belief that some species appeared spontaneously in nonliving matter.

**static electricity**   An imbalance of positive and negative charges.

**stomate**   The part of a plant located on the underside of leaves which releases unnecessary water that has been passed to the leaf of the plant through the stem.

**subject site**   A collection of human-compiled websites that have been gathered and organized according to their subject matter.

**supersaturated**   The state of a solution when excessive solute has been dissolved by heating.

**transpiration**   A process in which plants release unnecessary water through their leaves.

**unsaturated**   The state of a solution in which there is ample space between the molecules of the solvent.

**variable**   A single factor in an experiment that is intentionally altered, thereby changing the outcome of the experiment.

**virtual science fair**   An online science fair in which students are presented with real-life problems and asked to solve them using available technology. Virtual fairs are submitted and judged electronically.

**windchill factor**   The temperature of windless air that would have the same effect on exposed human skin as a given combination of wind and air temperature.

# Additional Resources

There are lots of good websites and books concerning science and science fair projects, and the more you read and learn, the more interesting science will be to you.

Check out some of the following websites listed to learn more about different areas of science and to get some more great ideas for projects.

## Not-to-Miss Science Websites

"The Earth and Moon View" contains a map of the earth that you can view from the sun, moon, a satellite in orbit, or from locations you specify using longitude and latitude. Find it at www.fourmilab.ch/earthview/vplanet. html.

Sponsored by NASA, the Liftoff to Space Exploration site contains space science and space-related information especially geared toward teens. It's on the web at liftoff.msfc.nasa.gov/.

The main NASA site, located on the web at www.gsfc.nasa.gov/, contains links to photos from space and allows you to view a shuttle launch.

"NASA Kids" has a variety of games and projects designed to get children of all ages excited about learning about space. It's at kids.msfc.nasa.gov/.

"Physics Life" is an interactive site relating physics to areas that kids relate to, such as playground activities. You can go there at www.physics.org/life.

"The Periodic Table of the Elements on the Internet" provides detailed information about the periodic table, as well as links to other sites. Find it at chemicalelements. com.

"Chem4Kids" has biographies of famous chemists, chemistry quizzes, and good information concerning atoms, elements, and matter. You'll find it at www.chem4kids. com/.

"A Guided Tour of the Visible Human" provides an animated look at the human body and how it works. It's at www.madsci.org/~lynn/VH/.

Intended for middle school ages and up, Frogguts.com contains a step-by-step, interactive frog dissection. Find it at www.froguts.com.

Sponsored by The Oceanic Group, "Wonders of the Seas" highlights different forms of marine life. It's on the web at www.oceanicresearch.org/lesson.html.

The "Bill Nye the Science Guy" website includes a lot of general science information, and allows you to email your science questions. It's located at www.billnye.com.

"Cool Science for Curious Kids" was created by the Howard Hughes Medical Institute, a nonprofit medical research organization based in Chevy Chase, Maryland. It offers online and offline science activities. You can find it at www.hhmi/org/ coolscience/.

An extensive listing of information pertaining to the scientist and his theories, "Albert Einstein Online" also contains many links to other sites, plus biographies and photographs of Einstein. It's at www.westegg.com/einstein.

"How Stuff Works" explains the mechanisms of everything from digital cameras to identity theft. It's on the web at www.howstuffworks.com.

Oceanography and space science are the main focuses of "The Office of Naval Research and Technology Focus," located at www.onr.navy.mil/focus/.

"A Science Fair Project Resource Guide," sponsored by the Internet Public Library, has great links to other science fair sites. You can get there at www.ipl.org/youth/ projectguide.

Ideas for science fair projects and all kinds of information about earthquakes and how to predict them are included on "Earthquakes for Kids & Grownups," a site sponsored by the U.S. Geological Survey. You can access the site at earthquake.usgs.gov/ 4kids/.

Environmental issues, organic farming, and land use can be explored on "Kids Re-Generation Resource Network," a website that also includes crafts and games. It's on the web at www.kidsregen.org.

"Building Big," allows kids to explore construction and participate in virtual labs. It includes information about various types of engineering jobs, and can be found at www.pbs.org.wgbh/buildingbig/.

The science of hockey, tracking severe storms and how to dissect a cow's eye are some topics to learn about on "The Exploratorium" website. It's sponsored by the science museum in San Francisco, and you can find it at www.exploratorium.edu/.

# Books

Listed below are some good books to provide additional science fair ideas and information, or just to teach you more about science. All the books can be found online at www.amazon.com or www.bn.com, or may be available in your local bookstore.

Atkins, Peter William. *A Journey into the Land of the Chemical Elements*. Cambridge, MA: Perseus Publishing, 1997.

Bochinski, Julianne Blair. *The Complete Handbook of Science Fair Projects*. Hoboken, NJ: John Wiley & Sons, Incorporated, 1996.

Bosak, Susan V., *A Source Book of Fascinating Facts, Projects and Activities*. Ontario: Scholastic Canada, Ltd., 2000.

Churchill, E. Richard. *365 Simple Science Experiments with Everyday Materials*. New York: Black Dog & Leventhal Publishers, Inc., 1997.

Downie, Neil A. *Vaccum Bazookas, Electric Rainbow Jelly, and 27 Other Saturday Science Projects*. Princeton, NJ: Princeton University Press, 2001.

Ehrlich, Robert. *Turning the World Inside Out and 174 Other Simple Physics Demonstrations*. Princeton, NJ: Princeton University Press, 1990.

Glandon, Shan. *Caldecott Connections to Science*. Westport, CN: Greenwood Publishing Group, Inc., 2000.

Hirschmann, Kris. *101 Science Fair Projects*. Mahwah, NJ: Troll Communications L.L.C., 2001

Jones, Charlotte Foltz. *Mistakes That Worked*. New York: Random House Children's Books, 1994.

Kohl, MaryAnn F., *Science Arts: Discovering Science Through Art Experiences.* Bellingham, WA: Bright Ring Publishing, Inc., 1993.

Krieger, Melanie Jacobs. *How to Excel in Science Competitions.* Berkley Heights, NJ: Enslow Publishers, Inc., 1999.

Moje, Steven W. *Cool Chemistry: Great Experiments With Simple Stuff.* New York: Sterling Publishing Company, Inc., 2001.

Rhatigan, Joe and Heather Smith. *Sure-to-Win Science Fair Projects.* New York: Sterling Publishing Company, Inc., 2002.

Robinson, Tom Mark. *Everything Kids' Science Experiments Book: Boil Ice, Float Water, Measure Gravity—Challenge the World around You!* Avon, MA: Adams Media Corporation, 2001.

St. George, Judith. *So, You Want to Be an Inventor?* New York: Penguin Putnam Books for Young Readers, 2002.

Sobey, Ed J. *How to Build Your Own Prize-Winning Robot.* Berkley Heights, NJ: Enslow Publishers, Inc., 2002.

Tocci, Salvatore, *Experiments With Electricity.* New York: Scholastic Library Publishing, 2002.

———. *Experiments With Light.* New York: Scholastic Library Publishing, 2002.

———. *Experiments With Magnets.* New York: Scholastic Library Publishing, 2001.

VanCleave, Janice. *Janice VanCleave's Guide to the Best Science Fair Projects.* Hoboken, NJ: John Wiley & Sons, Incorporated, 2000.

———. *Janice VanCleave's Help! My Science Project is Due Tomorrow! Easy Projects You Can Do Overnight.* Hoboken, NJ: John Wiley & Sons, Incorporated, 2001.

Vecchione, Glen. *100 First-Prize Make-It-Yourself Science Fair Projects.* New York: Sterling Publishing Company, Inc., 1999.

# Appendix C

# Sources for Scientific Supplies

Although most of the projects in this book don't require special items, a few of them may call for materials you can't find in your house or buy in an electronics or home supply store.

Following are some companies from which you can order scientific supplies. Even if you don't need to order anything for your current science fair project, you might want to keep these names handy for future projects.

Remember, though, to check first with your teacher about supplies you could borrow for your experiment and then return. Even if your teacher doesn't have a particular item that you need, he or she may know someone who does.

- **American Science Surplus/Jerryco.** This company offers surplus equipment at discounted prices, along with electronics and other supplies. You can get a catalog of items by contacting the company at PO Box 1030, Skokie, IL 60076, or calling 847-647-0010. The fax number is 1-800-934-0722, and the website is www.sciplus.com.

- **Carolina Biological Supply Company.** This is probably the largest and best-known house for science supplies. It's on the web at www.carolina.com, or you can contact the company at 2700 York

Road, Burlington, NC 27215-3398. The toll-free phone number is 1-800-334-5551 and the fax number is 1-800-222-7112.

◆ **Edmund Scientific.** This company offers science educational supplies, as well as some surplus. It's on the web at www.edsci.com. Or contact the company at 101 E. Gloucester Pike, Barrington, NJ 00807-1380. The phone number is 609-573-6250.

◆ **The Electronic Goldmine.** This supplier offers an extensive catalog of goods, along with electronic parts, solar cells, and so forth. Contact it at PO Box 5408, Scottsdale, AZ 85261. The phone number is 602-451-7454, and the fax number is 602-451-9495.

◆ **MWK Industries.** This company specializes in laser equipment and is known for its great prices. It's on the web at www.mwkindustries.com, or you can contact it at 1269 W. Pomona, No.112, Corona, CA 91720. The phone number is 909-278-0563 and the fax number is 1-909-278-4887.

◆ **School Aids.** This site is geared toward teachers and parents, and has an online catalog filled with useful materials for students embarking on science fair projects. Check it out at edumart.com/SA_SSCG/. You also can contact the company at 9335 Interline Ave., Baton Rouge, LA 70809. The phone number is 225-923-0294, and the fax is 225-923-1650.

# Index

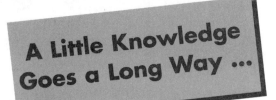
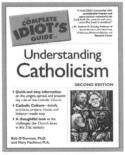

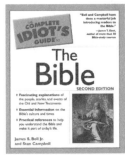

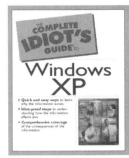